Springer Theses

Recognizing Outstanding Ph.D. Research

For further volumes:
http://www.springer.com/series/8790

Aims and Scope

The series "Springer Theses" brings together a selection of the very best Ph.D. theses from around the world and across the physical sciences. Nominated and endorsed by two recognized specialists, each published volume has been selected for its scientific excellence and the high impact of its contents for the pertinent field of research. For greater accessibility to nonspecialists, the published versions include an extended introduction, as well as a foreword by the student's supervisor explaining the special relevance of the work for the field. As a whole, the series will provide a valuable resource both for newcomers to the research fields described, and for other scientists seeking detailed background information on special questions. Finally, it provides an accredited documentation of the valuable contributions made by today's younger generation of scientists.

Theses are accepted into the series by invited nomination only and must fulfill all of the following criteria

- They must be written in good English.
- The topic should fall within the confines of Chemistry, Physics, Earth Sciences, Engineering and related interdisciplinary fields such as Materials, Nanoscience, Chemical Engineering, Complex Systems and Biophysics.
- The work reported in the thesis must represent a significant scientific advance.
- If the thesis includes previously published material, permission to reproduce this must be gained from the respective copyright holder.
- They must have been examined and passed during the 12 months prior to nomination.
- Each thesis should include a foreword by the supervisor outlining the significance of its content.
- The theses should have a clearly defined structure including an introduction accessible to scientists not expert in that particular field.

Evgeny Barkhudarov

Renormalization Group Analysis of Equilibrium and Non-equilibrium Charged Systems

Doctoral Thesis accepted by
Imperial College London, UK

Springer

Author
Dr. Evgeny Barkhudarov
The Blackett Laboratory
Imperial College London
London
UK

Supervisor
Prof. Dimitri D. Vvedensky
The Blackett Laboratory
Imperial College London
London
UK

ISSN 2190-5053 ISSN 2190-5061 (electronic)
ISBN 978-3-319-06153-5 ISBN 978-3-319-06154-2 (eBook)
DOI 10.1007/978-3-319-06154-2
Springer Cham Heidelberg New York Dordrecht London

Library of Congress Control Number: 2014937957

Printed on acid-free paper

Springer is part of Springer Science+Business Media (www.springer.com)

Supervisor's Foreword

This thesis has two parts, each based on an application of the Wilson renormalization group (RG). In the form developed by Wilson in the 1970s, the RG provides the means for determining various quantities that characterize systems spanning many length and time scales. Originally used for the calculation of critical exponents and equations of state in critical phenomena, applications of the RG have broadened to many other branches of physics, including dynamical systems and chaos, fractals, and disordered systems, and has become a standard topic in the undergraduate curriculum.

A RG transformation is comprised of two steps: coarse graining, based on the systematic elimination of short wavelength fluctuations, followed by a rescaling to restore the system to its original size. The result of the RG transformation is a similar system, but with "renormalized" coupling constants, whose changes can be represented by a trajectory as these transformations are repeated. Fixed points of these trajectories are of particular interest, as they represent scale-invariant states of the system. There are three types of variables associated with transformations near fixed points: relevant, which are driven away from the fixed point and represent experimental quantities which must be set to their critical values, irrelevant, which are driven toward the fixed point, and marginal, which are neither relevant nor irrelevant. Marginal variables, though comparatively rare, play an important role in the RG analysis in the second part of this thesis.

There are various formulations of the Wilson RG, including closed-form differential generators and diagrammatic expansions. Generators can be expanded to obtain the terms that contribute to the RG transformation, while diagrammatic methods are useful for identifying such terms directly. Part I of this thesis is an RG analysis of the d-dimensional Coulomb gas, a classic problem that has been studied from many perspectives. The analysis carried out in this thesis is based on the generator formulation of the Wilson RG. This necessitates a functional representation of the thermodynamic behavior of the system, in this case, the grand canonical partition function, which was derived from a regularized model of charges placed on a hypercubic lattice. The goal of this analysis was (1) to determine the trajectories of the coupling constants under the iterative application of the RG, and therefore (2) to see if the Wilson RG could provide input into particle-in-cell simulations in plasma physics, which are the main family of simulation methods used in this field. The role of the RG was to identify the effect of coarse graining on the coupling constants as a function of the infrared and ultraviolet cutoffs, which are the lattice spacing and system volume, respectively. The RG calculations reproduced established results, but in a more concise and transparent form, motivated a somewhat generalized form of the original functional, and determined closed-form expressions for the Debye screening length as functions of the cutoffs.

The main part of this thesis is the application of the dynamical RG to magnetohydrodynamic (MHD) turbulence. Turbulence is a widespread phenomenon in the dynamics of fluid flow which is often studied through the Navier–Stokes equation for electrically neutral fluids. The onset of turbulence is characterized by the Reynolds number $\mathscr{R}$, a dimensionless constant expressed as the ratio of inertial forces which act on the fluid to the viscous forces responsible for energy dissipation from the fluid. When the viscous forces are much smaller than the inertial forces, i.e., $\mathscr{R} \gg 1$, large-scale excitations are formed in the fluid, which then cascade to ever smaller excitations to the scale where the energy is dissipated.

This cascade is a characteristic feature of turbulence. MHD turbulence is characterized by non-linear interactions between fluctuations of the magnetic field and the flow velocity over a range of spatial and temporal scales, and plays an important role in several plasma phenomena, such as heating, the transport of energetic particles, and radiative transfer. MHD turbulence is also a key aspect of space and astrophysical plasmas.

The basic RG procedure is the same as for the d-dimensional Coulomb gas, but the evaluation of the pertinent terms is more involved because of the vector nature of the equations, which are an amalgamation of the Navier–Stokes and Maxwell equations, and the presence of stochastic stirring terms. The properties of quasi-neutral fluids governed by stochastic MHD equations are investigated with a method based on the diagrammatic representation of the Martin–Siggia–Rose field-theoretic formulation of stochastic dynamics. After the transformation to Elsasser variables, which is a symmetrization of the original equations, the solution is represented as a functional integral. The Jacobian resulting from the change of field variables was shown to reduce to a field-independent constant, which could, therefore, be omitted from the analysis. The coarse graining of the functional integral—involving a heroic bout of algebra—and its subsequent rescaling were cast differential equations for the analysis of RG trajectories.

The RG calculation identified an infinite set of diagrams which are marginal in the RG sense, that is, they neither grow nor decay under the action of the RG, as described above. The same problem arises for the randomly stirred Navier–Stokes equation. The marginal variables can be suppressed by working near equilibrium, where stochastic forcing results from thermal fluctuations, which enables the imposition of the fluctuation-dissipation theorem. Although such restrictions mean that the couplings in the Navier–Stokes equation remain at their microscopic values, for MHD, the effective diffusivity and viscosity tend to non-trivial values. Most importantly, these calculations demonstrate that, within this framework, marginal terms can be ignored, together with all higher-order corrections.

A non-zero correction to the three-point vertex was also computed, which introduces new structure: the kinetic equation is renormalized, while the induction equation is not. A consistent methodology for self-checks at this order would require the Ward–Takahashi identities, which provide relations between three-point vertices. This does not appear to have been reported in the literature. Since there is no small expansion parameter in the RG analysis, the effect of higher order terms, which need not be small, must be thoroughly analyzed near a fixed point before any physical predictions can be made. At least in the case of equilibrium fluctuations, such an analysis seems to be possible. An analysis with a more general spectrum for a more realistic model of turbulence is of paramount importance to justify the use of the RG.

A consistent description of MHD far from equilibrium is still absent, but this thesis highlights several aspects of the functional integral formulation with regard to the symmetries of the system and proposes possible ways to study this system non-perturbatively.

London, March 2014 Prof. Dimitri D. Vvedensky

Abstract

In this thesis we investigate properties of equilibrium and non-equilibrium systems by means of renormalization group (RG) analysis. In the study of the d-dimensional Coulomb gas, we have formulated a continuum model from the underlying hypercubic lattice and employed the irreducible differential formulation of the Wilson RG. We have identified a Thouless–Kosterletz transition in $d = 2$ and found no non-trivial fixed points for $d > 2$. As an example of a non-equilibrium system, we have investigated properties of quasi-neutral plasmas which are governed by stochastic magnetohydrodynamic (MHD) equations. The present method is based upon the Martin–Siggia–Rose field-theory formulation of stochastic dynamics. We develop a diagrammatic representation for the theory and carry out a momentum-shell RG of Wilson–Kadanoff type. An infinite set of diagrams is identified which are marginal in the RG sense. We have shown, in accordance with the previous literature, that the same problem arises for the randomly forced Navier–Stokes equation. The problem of marginal variables can be suppressed by working near equilibrium, where stochastic forcing represents thermal fluctuations. In a similar manner, we have considered regimes when MHD equations are subject either to kinetic or magnetic forcing only. In such models, the macroscopic limit can be taken such that all marginal terms are irrelevant and the dynamics is governed by linear equations. Furthermore, non-trivial fixed points are identified in such regimes and limiting values of either kinematic viscosity or magnetic diffusivity are derived. A consistent description of MHD dynamics far from equilibrium is still absent. We highlight some of the aspects of the functional integral formulation with regard to the symmetries of the system and propose possible ways in which the system can be studied non-perturbatively.

Acknowledgments

Most of all, I would like to thank my supervisor Prof. D. D. Vvedensky for his continuous help and support throughout my studies and to dedicate this work to my parents.

Contents

Chapter 1
Introduction

1.1 Historical Overview

1.1.1 Quantum Field Theory

The search for unifying principles of quantum mechanics with special theory of relativity began at the end of the 1920s. However, it had been realized from the beginning that relativistic quantum field theory was plagued by the insuperable difficulty of the ultraviolet (UV) divergences. As is now well known, the search for a solution to the problem of subtraction of UV divergences, lasting from the 1940s to the 1970s, led to the establishment of a consistent theoretical and mathematical formulation called the renormalization group (RG). The idea of the RG was originally formulated by Stuckelberg and Peterman in 1953 [1], and independently by Gell-Mann and Low in 1954 [2]. At the beginning of the 1970s a number of UV subtraction schemes were available and were proven to be equivalent. As such the RG expressed the invariance under different procedures of making the theory finite.

The discovery of the Dirac equation in 1928 has allowed a more precise description of electromagnetic interactions with charge particles. This framework, which consists of quantum and relativistic mechanics to describe such interactions, is known as quantum electrodynamics (QED). At the time, it was known that the rest mass of an electron in the classical relativistic framework, which is treated as a sphere of radius R, was divergent in the point particle limit. As such, it was hoped that quantum mechanics would reconcile the problem. In 1934 Weisskopf carried out a first consistent calculation which established that a contribution to the mass of the electron is also divergent (logarithmically). This was attributed to interactions with virtual photons of arbitrarily high momentum due to the absence of the small scale cut-off in QED. It was understood that these infinities are of profound importance and form the essence of the theory. Subsequent research into the nature of infinities established that logarithmic divergences persisted in the calculation of other physical quantities. It was noticed that by forming particular combinations of such physical

E. Barkhudarov, *Renormalization Group Analysis of Equilibrium and Non-equilibrium Charged Systems*, Springer Theses,
DOI: 10.1007/978-3-319-06154-2_1, © Springer International Publishing Switzerland 2014

quantities the infinities would cancel. The lack of a deeper understanding into the problem made it difficult to interpret these results.

Following experimental measurements of the Lamb shift in 1947 by Lamb and Retherford [3] and subsequent theoretical calculation, which was in a good agreement with the experiment, reassured the methodology behind reshuffling the infinities between physical quantities. The general idea of the method is the following. The initial theory is characterized by *bare* parameters such as electron mass and charge, $\{m_0, e_0\}$. One then introduces a momentum cut-off, which renders the integrals finite but otherwise arbitrary. Then, the observed quantities are calculated as a function of the bare parameters and the cut-off to a given order. If one then inverts the resultant relations to express the bare quantities as a function of the cut-off and the observed parameters, then upon substitution of those into observable/measurable quantities, the infinities which are carried by the cut-off cancel. Most importantly, these quantities are then insensitive to the cut-off, so by taking the limit $\Lambda \to \infty$, they remain finite.

This procedure led to remarkable agreements with the experiment, but it lacked a physical interpretation. One of the attempts to justify renormalization was given by Bogoliubov and others [4], which linked the problem of infinities in the bare parameters to a mathematical problem of a correct definition of singular products of distributions. A more physical interpretation consisted of the fact that the introduction of Λ meant that quantum field theories give a correct picture up to a given scale, beyond which some unknown physics take place. Renormalizable theories then were thought off as those which are insensitive to this unknown behavior.

In 1950s it was noted that massless theories have another peculiar property. In massive theories electric interactions can be defined through the Coulomb force between particles at rest. In massless theories, such a definition is no longer feasible and one seeks to introduce an arbitrary energy (mass) scale with respect to which one can define renormalized charge. Since this scale is arbitrary one can define another scale and, hence, an effective charge which would give the same physical results. The transformation from one set of parameters to another, which would not alter the physics, were then called the RG transformation.

Renormilizability gradually emerged as new law of nature, namely that all physical theories are renormalizable. Application of RG methods led to the triumph of quantum field theory. Subsequent development in non-Abelian gauge theories led to the construction of theories of combined weak and electromagnetic interactions. In 1973 Wilczek and Politzer [5–7] discovered asymptotic freedom using RG methods. Thus, by the mid-1970s a complete quantum field theory (QFT) description of all fundamental forces except gravity was proposed, which formed the Standard Model. The RG was an indispensable tool in this development.

1.1.2 Statistical Mechanics

In the context of critical phenomena the RG was developed as a tool to understand systems in which fluctuations persist to a macroscopic scale. The theory of critical phenomena primarily deals with second order-phase transitions. Examples include

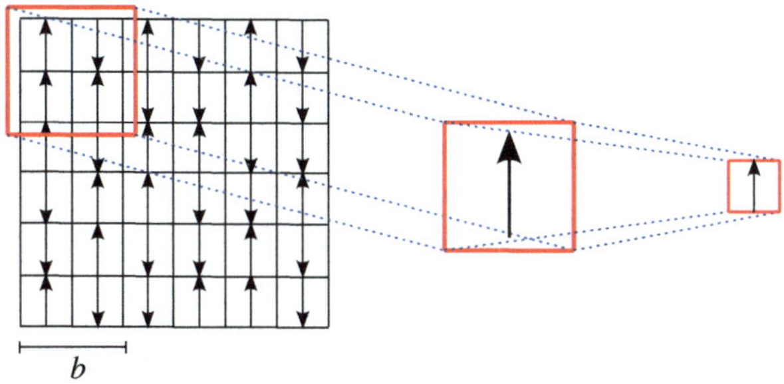

Fig. 1.1 The spins are correlated over a finite length, the correlation length. Kadanoff argued that by regarding a set of spins within the region of the correlation length as a single block spin, it may be a plausible to derive an effective theory of long interactions. This marked the beginning of the renormalization group in the context of statistical mechanics. As depicted in the diagram one begins by dividing the lattice into blocks with spacing b. By prescribing some sort of coarse-graining rule, the individual blocks of size b^d are assigned a spin. Since the lattice spacing now differs from the original lattice all linear scales are reduced by a factor of b. One then obtains are coarse-grained version of the original model

liquid-vapor or magnetic transitions. These phenomena fall into a class of problems where fluctuations penetrate all length scales of the system. Such problems are difficult to treat mathematically as there are many coupled degrees of freedom. The first attempts to understand such phenomena relied on the assumption that the system can be described by the means of a few macroscopic variables and the microscopic structure, such as atom spacings and the range of interactions, are not relevant for the macroscopic description. Intuitively this means that physics on very different scales decouples. As an example consider classical mechanics, where it is implicitly assumed that the atomic level interactions are irrelevant for, say, a description of terrestrial body dynamics. These considerations lead to the mean field theory description of critical phenomena (Fig. 1.1).

A characteristic feature of second order phase transition is that near the transition quantities of interest, such as the correlation length, diverge. Other physical quantities of interest either vanish or diverge as one approaches the transition point. The success of the mean field theory (MFT) description was seriously questioned by numerical and experimental studies. It failed to correctly predict the exponential behavior of physical quantities near the transition. The concept of universality predicted by MFT, namely that quantities diverge by the same exponents, regardless of the microscopic picture, still survived, but in a more limited sense. Universality properties seemed to depend on dimensions and symmetries of the system.

Such phenomena, namely, the coupling of different scales was new to physicists at the time and posed a great conceptual difficulty. Predictions of MFT have shown that there are systems where such decoupling of degrees of freedom is not always appropriate. Thus, it seems that the problem of infinities, which were first encountered in QFT found their way back into the theory of critical phenomena. Recall that, in QFT, it is the microscopic scale which when send to zero caused such divergences, whereas in critical phenomena the problem is due to the fact that a macroscopic scale, namely the correlation length, went to infinity.

A conceptual break through was made by Kadanoff [8], which was later formalized by Wilson [9–12]. Kadanoff's formulation argued, in the context of the Ising model, that near criticality averaging over groups of spin (microscopic variables) assigns to the original system another corresponding physical system having the *same* behavior at long distances.

This transformation can then be iterated. If this procedure generates an effective interaction whose asymptotic form is independent of the initial microscopic behavior one then finds a mechanism to explain universality. This is manifested by the existence of fixed points of this iteration procedure. At this point we would like to spend some time to explain the Wilson-Kadanoff RG in some more detail, as it forms the basis of our analysis in this thesis.

1.2 Wilson-Kadanoff Renormalization Group

1.2.1 Definition of a Renormalization Group Transformation

The basic idea is to study large-scale properties of a theory by partial elimination of short-scale degrees of freedom. The change in the effective theory through elimination of degrees of freedom and subsequent rescaling can be visualized as a dynamic flow in the space of theories with a fixed cutoff. Scaling behavior can be identified with the existence of fixed points in the dynamic flow. The concept of universality is then understood from the properties of the manifold in the proximity of a fixed point. Our discussion will closely follow [13, 14] (Fig. 1.2).

In the context of dynamical problems, the RG acts on the space of dynamic probability distributions, which are characterized by the equations of the form:

$$\frac{\partial \mathbf{v}(\mathbf{x},t)}{\partial t} - K(\mathbf{v}) = \mathbf{f}(\mathbf{x},t), \tag{1.2.1}$$

where $\mathbf{v}$ is a physical field, $K(\cdot)$ is some functional of the physical field variable and $\mathbf{f}$ is the noise term. These are most conveniently described in a path-integral representation using the Martin-Siggia-Rose action [15]. For hydrodynamics driven by random noise, the generating functional is:

$$\mathscr{Z} = \int \mathscr{D}[\mathbf{v}]\,\mathscr{D}\left[\hat{\mathbf{v}}\right] e^{-S[\mathbf{v},\hat{\mathbf{v}},\Lambda]}, \tag{1.2.2}$$

where S is the action which depends on the physical field $\mathbf{v}$, the conjugate noise field $\hat{\mathbf{v}}$ and the cut-off Λ and has the following form:

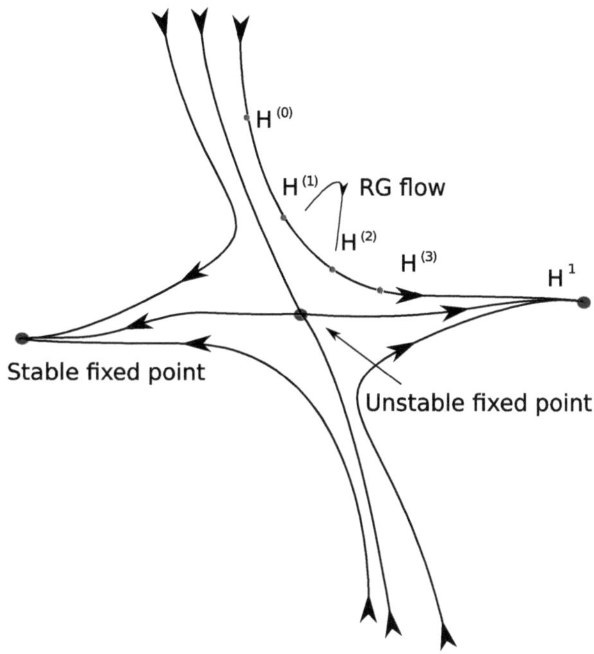

Fig. 1.2 The diagram depicts the flow of couplings in the parameter space under the action of RG transformation. The relevant variables are those which direct the flow away from a fixed point while irrelevant variables drive the flow towards a fixed point

$$S\left[\mathbf{v}, \hat{\mathbf{v}}\right] = \int d^d\mathbf{x}\, dt\, \hat{\mathbf{v}}\left(\mathbf{x}, t\right) \cdot F\left(\mathbf{x} - \mathbf{x}', t - t'\right) \cdot \hat{\mathbf{v}}\left(\mathbf{x}', t'\right)$$
$$+ i \int d^d\mathbf{x}\, dt\, \hat{\mathbf{v}}\left(\mathbf{x}, t\right) \cdot \left(\frac{\partial \mathbf{v}\left(\mathbf{x}, t\right)}{\partial t} - K\left(\mathbf{v}\right)\right). \tag{1.2.3}$$

In Fourier space, a cut-off is introduced to render the integrals finite in the UV region. Note that, in the context of critical phenomena, the existence of a cut-off is natural. Its origins can be traced to lattice models, which have an intrinsic minimum scale. The quantity $F\left(\mathbf{x} - \mathbf{x}'\right)$ carries the information about the noise statistics. The field variables are separated into a low and high wave number components:

$$\mathbf{v}\left(\mathbf{x}, t\right) = \mathbf{v}^{>}\left(\mathbf{x}, t\right) + \mathbf{v}^{<}\left(\mathbf{x}, t\right), \tag{1.2.4}$$

such that the field is projected onto $\left[\Lambda b^{-1}, \Lambda\right]$ and $\left[0, \Lambda b^{-1}\right]$ wave numbers, respectively. Projections onto $\left[\Lambda b^{-1}, \Lambda\right]$ represent microscopic degrees of freedom, which we seek to eliminate. The parameter b measures the degree of filtering, with $1 \leq b < \infty$. The probability distribution for the low-pass filtered fields remains unchanged by integrating out high-pass filtered fields, i.e. the large-scale properties of the systems are not altered. The generating functional for the effective theory then takes the form:

$$\mathscr{Z} = \int \mathscr{D}\left[\mathbf{v}^{<}\right] \mathscr{D}\left[\hat{\mathbf{v}}^{<}\right] e^{S_{\text{eff}}\left(\mathbf{v}^{<}, \hat{\mathbf{v}}^{<}\right)}. \tag{1.2.5}$$

In general, the form of $S_{\text{eff}}\left(\mathbf{v}^{<}, \hat{\mathbf{v}}^{<}\right)$ would be different and involve new interactions which were not present in the original formulation. The next step in the RG program is rescaling. We seek to restore the cut-off. The rescaling operation includes rescaling of both the space and time variables and the fields:

$$\mathbf{v}^{<}(\mathbf{x}, t) \rightarrow b^{\alpha}\mathbf{v}\left(b^{-1}\mathbf{x}, b^{-z}t\right). \tag{1.2.6}$$

The scaling exponents α and z are to be determined. Note that, the above operations do not alter the probability distribution, which describes large-scale properties of the theory. Although we have restored the cut-off to its original length, we have reduced the density of points in Fourier space, so the number of degrees of freedom in the effective action has been reduced. Collectively all these manipulations form the RG transformation, which can be visualized as a map which transforms one probability distribution into another, or equivalently one action into another:

$$\mathscr{R}_b : S \rightarrow S', \tag{1.2.7}$$

where S' is the effective action. As we have mentioned, the RG transformations can be visualized as a flow in the parameter space. Given the action, the probability distribution is completely specified by the coupling constants, which form a set of parameters μ:

$$\mu = (u_1, u_2, \ldots, u_n). \tag{1.2.8}$$

The RG transformation can be thought of as a transformation from one set of parameters to another:

$$\mathscr{R}_b\mu = \mu', \tag{1.2.9}$$

where μ' is a set of parameters which are coefficients in front of field variables in S'. The coefficients strongly depend on the cut-off, as it describes the scale to which a given description in terms of μ is valid. The necessity to restore the cut-off then becomes apparent as it allows direct comparison between the two sets μ and μ', since both are dependent on Λ.

So far, the elimination of small-scale degrees of freedom was considered over the $\left[\Lambda b^{-1}, \Lambda\right]$ shell in momentum space. By successively eliminating the degrees of freedom further, we observe that the RG transformation obeys the following rule (closure):

$$\mathscr{R}_b\left(\mathscr{R}_b\mu\right) = \mathscr{R}_{b^2}\mu, \tag{1.2.10}$$

which results in the peculiar form of field rescaling. However, since there is no inverse of the $\mathscr{R}_b$ it does not form a group but a semi-group. A fixed point under $\mathscr{R}_b$ in the parameter space satisfies:

$$\mathscr{R}_b \mu^* = \mu^*. \tag{1.2.11}$$

In general this would be a non-linear equation. However, if the presence of a fixed point can be established, the above equation can be linearized in its vicinity. Formally we can write $\mu = \mu^* + \delta\mu$. The RG operator can then be linearized if we ignore corrections of order $\mathscr{O}\left(\delta\mu^2\right)$ and higher. The fixed point equation reduces to an eigenvalue problem:

$$\mathscr{R}_b^L \delta\mu = \delta\mu'. \tag{1.2.12}$$

The solution can be written in its most general form as a linear combination of the eigenvectors:

$$\delta\mu' = \sum_j t_j b^{y_j} e_j, \tag{1.2.13}$$

where b^{y_j} is an eigenvalue of e_j. Depending on the sign of y_j, the variables, being either positive, negative or zero, Wilson classified them as relevant, irrelevant and marginal, respectively. A subspace $\mathscr{C}$, where $y_j < 0 \ \forall \ e_j \in \mathscr{C}$ defines a critical surface.

Points on $\mathscr{C}$ will be pushed towards the fixed point, while points defined not on the critical surface with $y_j > 0$ will drive away from the fixed point. In case of the marginal variables, when $y_j = 0$, the situation becomes more complicated since nothing can be inferred about the behavior of the variable in a linear approximation. As has been emphasized by Wilson [9–12], the terms which can be neglected in the RG analysis crucially depend on the irrelevancy of these variables. A resolution of such cases usually requires a higher-order analysis (e.g. higher-order loop expansion).

The quantities of interest such as critical exponents can be explicitly calculated in the vicinity of a fixed point. A concept of a manifold in the parameter space $\mathscr{C}$ allows one to introduce the idea of universality in a natural manner. Theories which share the same relevant variables near a fixed point will have the same large-scale properties. Therefore, all such theories will have the same critical exponents.

1.2.2 Renormalization Group in Field Theory and Statistical Mechanics

In statistical mechanics the divergence of the correlation length is responsible for the difficulty of analyzing the system near a phase-transition. To be more specific we consider a loop integral which one encounters in the ϕ^4 theory [16, 17]:

$$I_{SM} = \int_0^\Lambda \frac{d^d\mathbf{q}}{\xi^{-2} + q^2}. \tag{1.2.14}$$

At the critical point the correlation length goes to infinity, therefore the behavior of the integral I is completely determined by the dimensionality of the system:

$$\lim_{\xi\to\infty} I_{SM} = \begin{cases} d > 2 & \text{convergent,} \\ d = 2 & \text{logarithmically divergent,} \\ d < 2 & \text{divergent.} \end{cases} \tag{1.2.15}$$

The divergence comes from the infra-red (IR) region. For $d = 1$ we observe that the integral behaves as:

$$\lim_{\xi\to\infty} I_{SM} = \int_0^{\Lambda} q^{-2} dq \int d\Omega, \tag{1.2.16}$$

where $\int d\Omega$ is an integral over the surface of a unit sphere. Analogous integrals are encountered in QFT. However QFT is formulated to have the Lorentz metric so formally the integrals are analogous once QFT is re-written to have a Euclidean metric by means of a Wick rotation. More importantly the integral reads:

$$I_{QFT} = \lim_{\Lambda\to\infty} \int_0^{\Lambda} \frac{d^d\mathbf{q}}{m_0^2 + q^2}, \tag{1.2.17}$$

where m_0 is a particle's mass, which we assume is non-zero. The integral behaves as:

$$I_{QFT} = \begin{cases} d > 2 & \text{divergent,} \\ d = 2 & \text{logarithmically divergent,} \\ d < 2 & \text{convergent.} \end{cases} \tag{1.2.18}$$

The divergence comes from the UV region. The two integrals behave very differently. While in QFT it is the UV cut-off which is responsible for the divergence of the loop corrections in statistical mechanics the correlation length plays the role of the IR cut-off. In the first case the theory does not posses an intrinsic minimum scale so one naturally looks to send $\Lambda \to \infty$. On the contrary, in statistical mechanics the cut-off is natural but one is interested in properties of the system near a phase transition when the IR cut-off goes to zero. In that sense we can see that RG program implemented in QFT is rather different from the Wilson-Kadanoff scheme.

In this work we will be only concerned with lowering the cut-off, as opposed to increasing the cut-off, as one would do in QFT. Note that, in the RG procedure, the cut-off is lowered incrementally $\Lambda \to \Lambda b^{-1} \to \Lambda b^{-2} \to \cdots$, so the loop corrections which are generated:

$$I_{SM} = \int_{\Lambda b^{-1}}^{\Lambda} \frac{d^d\mathbf{q}}{\xi^{-2} + q^2} \to \cdots, \tag{1.2.19}$$

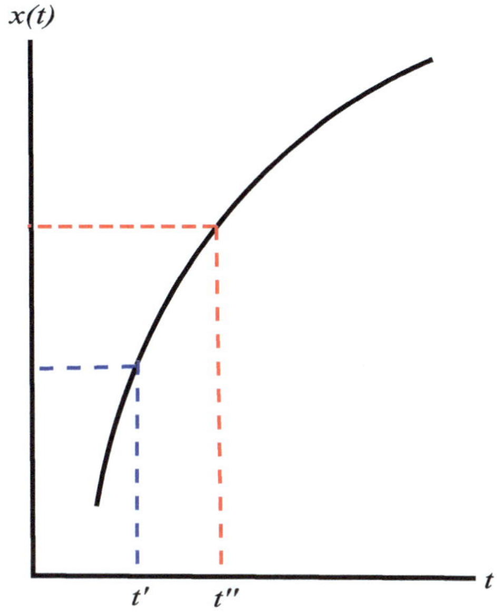

Fig. 1.3 A hypothetical solution to the differential equation is represented by a *black curve*. The value of the function, at some time $t > t'' > t'$, is independent of whether we choose to specify the function at t' or t''

are finite at every stage of the RG program. It follows that the divergence occurs as a result of the iteration of the RG. The study of such iterations of equilibrium and non-equilibrium models will be the main focus of this thesis.

1.2.3 Applications of Renormalization Group in Mathematical Physics

RG methods can be very useful in branches of physics other than high-energy physics and statistical mechanics. For instance, it can be used to solve partial differential equations (PDE) [18, 19]. Here, we would like to outline the general principle behind the method (Fig. 1.3).

Group theory has been an indispensable tool in analyzing symmetries of various PDE encountered in mathematical physics. If the equation admits a symmetry, together with the boundary conditions, the solution can be constructed in terms of the invariant of this symmetry group. Further, one can re-express the original PDE in terms of the canonical variables which can simplify the resultant equation to an ordinary differential equation. The foundations of this method have been laid by Sophus Lie and it is a well established methodology.

However, there exists another symmetry which is not of the actual equation but of the solution itself. Consider a transformation which leaves the solution invariant under a change in the initial condition. Such a symmetry is called a renormalization group. The mathematical aspects are very similar to those employed in QFT [20, 21]. The RG symmetry is an exact symmetry of the solution and some boundary values.

The difference between the blocking concept of Wislon-Kadanoff RG lies in the fact that the latter is an approximate semi-group.

The sole purpose of the method is to form additional differential equations which complement the original one. In the trivial case these additional equations simply reduce to the original differential equation and nothing can be inferred from the analysis. In non-trivial cases these additional RG equations are different and can be solved, providing a solution to the original equation. Let us illustrate the above concepts with a particular trivial example from [18]. Consider an equation:

$$\frac{dx}{dt} = V(x) . \qquad (1.2.20)$$

The solution is invariant under a time shift, so the solution will generally depend on two initial parameters t_0 and x_0. Thus, we can write:

$$x(t) = X(t - t_0, x_0) \quad \text{such that} \quad X(0, x_0) = x_0 . \qquad (1.2.21)$$

The last equation is simply the re-statement of the initial condition. Consider now that we want to choose an initial condition at some later instant in time $t_1 > t_0$ such that the initial position, x_1, lies on the same trajectory:

$$x_1 = X(t_1 - t_0, x_0) . \qquad (1.2.22)$$

Then, we can readily deduce the following:

$$x(t) = X(t - t_0, x_0) = X(t - t_1, x_1) , \qquad (1.2.23)$$

or more generally we can write ($t_0 = 0$):

$$X(t, x_0) = X(t - \tau,\ X(\tau, x_0)) . \qquad (1.2.24)$$

The above relation is a condition of functional self-similarity. The above transformation obeys the axioms of a group. Thus a transformation:

$$t \Rightarrow t_1 = t - \tau, \quad g \Rightarrow g_1 = f(\tau, g) \qquad (1.2.25)$$

obeys the group composition law:

$$g_2 = f\left(\tau + \tau', g\right) = f\left(\tau', g_1\right) . \qquad (1.2.26)$$

the identity element corresponds to $\tau = 0$ and the inverse by $-\tau$. However, one often encounters the multiplicative version of the above transformations by changing variables to $t = \ln x$ and $\tau = \ln \lambda$. As such, the statement of functional self-similarity reads:

$$F(x, g) = F\left(x\lambda^{-1}, F(\lambda, g)\right) . \qquad (1.2.27)$$

From the above equation one can derive a differential equation satisfied by $F(x, g)$:

$$\left\{-x\frac{\partial}{\partial x} + \beta(g)\frac{\partial}{\partial g}\right\} F(x, g) = 0, \tag{1.2.28}$$

which follows from taking a derivative of the function with respect to λ and then setting $\lambda = 1$. The β-function is:

$$\beta(g) = \left.\frac{\partial F(\lambda, g)}{\partial \lambda}\right|_{\lambda=1}. \tag{1.2.29}$$

The resulting equation is the well known Callan-Symanzik equation [22–24] derived in the QFT context, while in the Soviet literature the equation goes by the name of Ovsyannikov's compensation equation [25, 26]. The β-function, which is referred to as the Gell-Mann-Low or the Wilson function, plays a central role in the further analysis. Given the β-function the equation can be solved by the method of characteristics. If we parametrize the function $F(x(s), t(s))$ such that it stays invariant along the characteristic we have:

$$\frac{dF(s)}{ds} = \frac{\partial F}{\partial x}\frac{dx}{ds} + \frac{\partial F}{\partial g}\frac{dg}{ds} = 0. \tag{1.2.30}$$

It follows that along the characteristics we have:

$$\frac{dx}{ds} = -x, \tag{1.2.31}$$

$$\frac{dg}{ds} = \beta(g). \tag{1.2.32}$$

Thus, the characteristic has the form:

$$-\frac{dx}{x} = \frac{dg}{\beta(g)}. \tag{1.2.33}$$

Upon integration, this is the celebrated Gell-Mann-Low formula. The values of the β-function such that:

$$\beta(g_*) = 0, \tag{1.2.34}$$

form the basis of the search for the asymptotic solutions of the RG equation since the main contribution to the integral in the above equation comes from the region when $\beta(g_*) = 0$. Note that our derivation did not make any explicit reference to QFT as such, while it is precisely the machinery used in the field theory. In this sense RG has a much wider range of applications. To give an example, consider the following equation [19]:

$$\frac{\partial u(x, t)}{\partial t} = \sigma\frac{\partial}{\partial x}u^m(x, t)\frac{\partial u(x, t)}{\partial x}, \tag{1.2.35}$$

which is often encountered in mathematical physics. For instance, for $m > 1$ the equation is referred to as the equation of porous medium. From the symmetry argument of the differential equation it can be shown [19] that the solution has a general form:

$$u(x,t) = u_0 f\left(\frac{x}{\sqrt{\sigma\tau_0 u_0^m}}, \frac{t}{\tau_0}\right), \tag{1.2.36}$$

where $u_0 = u(x_0, t_0)$, τ_0 is a temporal time scale, which is introduced through a boundary condition $\partial_t u|_{(x_0,t_0)} = u_{0t}$. The property of functional similarity requires the following relationship to be satisfied:

$$u(x,t) = u_0 f\left(\frac{x}{\sqrt{\sigma\tau_0 u_0^m}}, \frac{t}{\tau_0}\right) = u_1 f\left(\frac{x-x_1}{\sqrt{\sigma\tau_1 u_1^m}}, \frac{t-t_1}{\tau_1}\right). \tag{1.2.37}$$

Through a number of algebraic manipulations and use of the boundary conditions, the requirement of functional self-similarity translates into the following functional relation:

$$f(\xi,\eta) = f(\xi_1,\eta_1) f\left((\xi-\xi_1)\sqrt{\phi(\xi_1,\eta_1) f^{-m}(\xi_1,\eta_1)}, (\eta-\eta_1)\,\phi(\xi_1,\eta_1)\right). \tag{1.2.38}$$

the parameters ξ_1 and η_1 play the role of the free parameter τ in our previous trivial example. The function ϕ is:

$$\phi = \frac{\partial \ln f}{\partial \eta}. \tag{1.2.39}$$

We do not intend to give full account of the above result but merely to illustrate the principle of the use of the RG in mathematical physics. The Gell-Mann-Low-like equations are derived by taking derivatives with respect to the free parameters ξ_1 and η_1. Since we have looked at the problem using additive properties rather than multiplicative, the parameters ξ_1 and η_1 are set to zero, not one. The resulting two differential equations are not equivalent to the original equation and form additional conditions which supplement the original PDE. We will not go into any more details but simply state the solution which follows from the solutions of the RG equations:

$$u(x,t) = u_0 (1+Bt)^b\, \psi\left(\sqrt{\frac{bB}{\sigma u_0^m}}x(1+Bt)^{-(mb+1)/2}\right), \tag{1.2.40}$$

where B and b are some parameters, which are specified by the boundary conditions.

This non-trivial result is a great example of how RG can be implemented in branches of physics other than QFT and SM. In addition it illustrates a deep mathematical concept of an exact symmetry of the solution with respect to the choice of the initial condition. We believe that the above presentation of the RG as it stands is more transparent than if one is to understand RG solely from the QFT perspective. In connection with the Kadanoff-Wilson picture we would like to paraphrase E. V. Teodorovich [18] "...one should not identify the RG method solely with Kadanoff's procedure for reducing the number of modes in a multi-mode system by sequential averaging ...". There are purely mathematical differences between the two which have to do with the group structure etc., and conceptually the self-similarity property is an exact symmetry of the solution while Wilson-Kadanoff RG is a transformation between various models which share the same large-scale properties.

References

1. E.C.G. Stueckelberg, A. Petermann, La normalization des constantes dans la theorie des quanta. Helv. Phys. Acta **26**, 499 (1953)
2. M. Gell-Mann, F.E. Low, Quantum electrodynamics at small distances. Phys. Rev. **95**, 1300 (1954)
3. W.E. Lamb, R.C. Retherford, Fine structure of the hydrogen atom by a microwave method. Phys. Rev. **72**(3), 241–243 (1947)
4. N.N. Bogoliubov, O.S. Parasyuk, A theory of multiplication of causative singular functions. Dokl. Akad. Nauk BSSR **100**, 25–28 (1955)
5. D.J. Gross, F. Wilczek, Ultraviolet behavior of non-abelian gauge theories. Phys. Rev. Lett. **30**, 1343–1346 (1973)
6. D.J. Gross, F. Wilczek, Asymptotically free gauge theories. I. Phys. Rev. D **8**, 3633–3652 (1973)
7. D.J. Gross, F. Wilczek, Asymptotically free gauge theories. II. Phys. Rev. D **9**, 980–993 (1974)
8. L.P. Kadanoff, Scaling laws for ising model near T_c. Physics **2**, 263 (1966)
9. K.G. Wilson, Renormalization group and critical phenomena. I. Renormalization group and the kadanoff scaling picture. Phys. Rev. B **4**, 3174 (1971)
10. K.G. Wilson, Renormalization group and critical phenomena. II. Phase-space cell analysis of critical behavior. Phys. Rev. B **4**, 3184–3205 (1971)
11. K.G. Wilson, The renormalization group: critical phenomena and the Kondo problem. Rev. Mod. Phys. **47**, 773–840 (1975)
12. K.G. Wilson, The renormalization group and critical phenomena. Rev. Mod. Phys. **55**, 583–600 (1983)
13. S. Ma, Introduction to the renormalization group. Rev. Mod. Phys. **45**, 589–614 (1973)
14. G.L. Eyink, The renormalization group method in statistical hydrodynamics. Phys. Fluids **6**, 3063 (1994)
15. P.C. Martin, E.D. Siggia, H.A. Rose, Statistical dynamics of classical systems. Phys. Rev. A **8**, 423 (1973)
16. D.J. Amit, *Field Theory, the Renormalization Group, and Critical Phenomena* (McGraw-Hill Inc., New York, 1978)
17. M. Le Bellac, *Quantum and Statistical Field Theory* (Oxford University Press, New York, 1991)
18. E.V. Teodorovich, The renormalization-group method in problems of mechanics. J. Appl. Math. Mech. **68**, 299–326 (2004)
19. E.V. Teodorovich, The renormalization-group approach to solving the equation of nonlinear transfer. J. Phys. A: Math. Theor. **42**, 155202 (2009)
20. V.F. Kovalev, D.V. Shirkov, Functional self-similarity and renormalization group symmetry in mathematical physics. Theor. Math. Phys. **121**, 1315–1332 (1999)
21. D.V. Shirkov, Renormalization group, invariance principle and functional self-similarity. Sov. Phys. Dokl. **27**, 197–200 (1982)
22. C.G. Callan Jr, Broken scale invariance in scalar field theory. Phys. Rev. D **2**, 1541–1547 (1970)

23. K. Symanzik, Small distance behaviour in field theory and power counting. Commun. Math. Phys. **18**, 227 (1970)
24. K. Symanzik, Small-distance-behaviour analysis and wilson expansions. Commun. Math. Phys. **23**, 49 (1971)
25. L.V. Ovsyannikov, Group properties of the equation of heat conductivity. Sov. Phys. Dokl. **125**, 492 (1959)
26. L.V. Ovsyannikov, *Group Analysis of Differential Equations* (Academic, New York, 1982)

Part I
Equilibrium Statistical Mechanics: Coulomb Gas

Chapter 2
d-Dimensional Coulomb Gas

2.1 Introduction

As was discussed in the previous chapter, renormalization group serves as a very powerfull tool in the study of strong interactions and fluctutations, which occur near continuous phase transitions. This powerful method presents a conceptual framework which allows for capturing complex natural phenomena in terms of a few mesoscopic variables. In this chapter we will consider an equilibrium system of charges, however, the concept can also be applied in the context of dynamic critical phenomena, which will be the subject of later chapters.

The d-dimensional Coulomb gas is a statistical mechanical problem where particles of equal or opposite charge interact through the Coulomb potential. The model has been extensively studied in the past and forms one of the classical problems in field theory [1].

To this end, most of the theoretical investigations have been centered around the particular case of two dimensions. In this regime the model undergoes the celebrated Kosterlitz-Thouless phase transition [2]. The integer-charged particles interact via a logarithmic potential. Alternatively, the charges can be viewed as vorticies which carry integer vorticity. At low temperature these vorticies are bound in pairs and carry zero vorticity and thus form an insulating state. At higher temperatures the binding of vorticies decreases until at some critical temperature, T_c, the vorticies are completely unbound, thereby forming a conducting state.

The special property of the Kosterlitz-Thouless phase transition is the behavior of the correlation functions. In the metallic state one observes screening because the charges are unbound. As a result, the correlation function decays exponentially fast. In the insulating phase the correlation function decays algebraically, meaning that charge fluctuations are correlated over infinite distances, hence the correlation length is divergent throughout the insulating phase. By contrast, in the Ising model and other second-order phase transitions, the correlation function, above and below some critical temperature, T_c, decays exponentially fast and only at the critical temperature

E. Barkhudarov, *Renormalization Group Analysis of Equilibrium and Non-equilibrium Charged Systems*, Springer Theses,
DOI: 10.1007/978-3-319-06154-2_2, © Springer International Publishing Switzerland 2014

the correlation functions decays algebraically. For that reason the insulating phase is referred to as a phase of quasi-long-range order.

The two-dimensional Coulomb gas falls into the same universality class as other statistical mechanical models such as the XY-model and the two-dimensional sine-Gordon model [1, 3]. Such mappings provide a relationship between seemingly different physical systems and provide an efficient tool for analyzing two dimensional problems in statistical mechanics [4, 5].

A part from the connection of the Coulomb gas to other physical systems, the d-dimensional Coulomb gas can be used as a model of a classical one-component plasma [5, 6]. Such plasmas are found in astrophysical systems such as white dwarfs, for example. In this thesis we will be primarily concerned with the plasma characteristics of the Coulomb gas rather than its connections to other physical models. Although the system does not undergo a phase transition for spatial dimensions $d > 2$, a renormalization group analysis can be utilized to study the role of the intrinsic length scales of the system in the behavior physical quantities such as the Debye length. Such calculations can be important in numerical simulations of plasma [7]. In the next section we will give an overview of renormalization group (RG) techniques used so far in the study of the Coulomb gas, before we present the Coulomb gas model formally and analyze it using a somewhat simpler RG method.

2.1.1 Renormalization Group Methodologies

If we go back to the particular case of $d = 2$, it must be stressed that while the argument in favor of Kosterlitz-Thouless phase transition can be relatively easily demonstrated by means of a perturbative RG analysis, the extrapolation of the behavior of the correlation function to the whole phase-space has proved to be a difficult task [1]. So, here we acknowledge some of the rigorous results which have been established for the $d = 2$ Coulomb gas [8]. Consider the metal-insulator phase space diagram:

Consider the red region in the Fig. 2.1, which represents the region of phase space sufficiently far from the critical line. It has been shown by [9] that the correlation function does indeed follow a power law decay for any z. For the opposite region, shaded blue, which is also sufficiently far from the critical line a somewhat similar result, but slightly weaker, consists of a proof that the correlation function decays exponentially for $|z| > 0$, subject to a specific boundary condition [10]. A number of other results which are concerned with rigorous proofs on the behavior of the correlation functions in different regions of the phase-space can be found here [8, 9, 11]. In this thesis we will not be concerned with these matters any further. Our analysis will be limited strictly to the vicinity of critical points.

As we have established in the first part of the thesis, there are a number of techniques to execute the RG program. One by means of field theory methods and the other by using the Wilson-Kadanoff style of RG. An extensive RG study of the Coulomb gas in two dimensions using the formalism of field theory is given by Amit [12]. In their detailed chapter the authors computed higher-order corrections in the

Fig. 2.1 Phase-space diagram of the $d = 2$ Coulomb gas. The two parameters of the system are z, the fugacity, and β, the inverse of the temperature. The *thick line* represents a critical line. To the *right* of the critical line, the low temperature regime, the system is in an insulating state. Vorticies are bound together, with the net vorticity being zero. To the *left* of the critical line charges are unbound which forms a metallic state

flow equations, to those originally derived by Kosterlitz [2], and have shown that vorticies with multiple charges are irrelevant in the RG sense. This investigations did not consider any other dimensions other than $d = 2$.

A version of real-space renormalization group RG has been given by [13]. In this chapter the RG flow has been derived for all physical dimensions. The coarse-graining of the action in the momentum space has received most of the attention in the literature. A paper exploiting the Wegner-Houghton [14] approach, for example, is [15]. A detailed calculation which exploits a standard cumulant expansion is [16].

In this thesis we present yet another RG scheme, which, to our knowledge, has not been used in the literature in the context of d-dimensional Coulomb gas. It is based on irreducible differential formulation of the Wilson-Kadanoff RG and has been successfully applied in the past to other models in statistical mechanics [17]. As suggested by the word irreducible, the object of interest is the vertex function or the Legendre transform of the free energy. The RG is formulated in terms of the integro-differential equation which describes the infinitesimal change of the vertex function under sequential mode averaging. Once compared to earlier mentioned calculations, we believe that this approach has distinct advantages in terms of conciseness and simplicity of the resultant integrals.

In the next section we will introduce the Coulomb gas model formally and derive the continuum theory in order to facilitate the Wilson-Kadanoff style of RG.

2.2 Functional Integral

2.2.1 Lattice Model

We begin by considering a d-dimensional Cartesian lattice whose spacing is a and with a total volume of V. Charges $\pm$e can occupy the sites i_k of this lattice and there is a Coulomb interaction between these charges. For $d = 3$, the lattice Coulomb potential between sites i_j and i_k is

$$U_{ij} = \frac{1}{4\pi a|i_j - i_k|}. \tag{2.2.1}$$

There are two main reasons for putting the system on a lattice. First, the lattice avoids the singularity of the Coulomb potential at the origin, which would otherwise necessitate introducing a short-range cut-off. This issue arises when we take the continuum limit, but for the moment there are no short-range singularities. The other advantage of a lattice formulation is that we can develop a field-theoretic formulation for the partition function, which facilitates the identification with mean-field limits and perturbations therefrom.

The canonical partition function $\mathscr{Z}_N$ for N charges $q_k = \pm\mathrm{e}$ at positions i_k, for $k = 1, \dots, N$, is

$$\mathscr{Z}_N = \sum_{\substack{\{i_k\} \\ \{q_k=\pm\mathrm{e}\}}} \frac{1}{N!} \exp\left[-\frac{\beta}{2} \sideset{}{'}\sum_{\substack{j,k \\ 1\le j,k\le N}} q_j U(i_j, i_k) q_k\right], \tag{2.2.2}$$

where the factor $N!$ is to ensure correct Boltzmann counting, $\beta = 1/k_B T$, k_B is Boltzmann's constant, T is the absolute temperature, the prime on the summation indicates that the terms $j = k$ are excluded, and the factor of $1/2$ is to avoid double counting. The associated grand canonical partition function Ξ is

$$\Xi = \sum_{N=0}^{\infty} z^N a^{Nd} \mathscr{Z}_N, \tag{2.2.3}$$

where $z = \mathrm{e}^{\beta\mu}$ is the fugacity and μ is the chemical potential.

The evaluation of Ξ proceeds by using the Hubbard–Stratonovich transformation, which is a standard operating procedure for such Hamiltonians. Based on the identity

$$\exp\left(\frac{b^2 y^2}{2a}\right) = \left(\frac{a}{2\pi}\right)^{1/2} \int_{-\infty}^{\infty} \mathrm{e}^{-\frac{1}{2}ax^2 + bxy}\, dx, \tag{2.2.4}$$

the Hubbard-Stratonovich transformation for a symmetric $N \times N$ matrix A with real eigenvalues is

$$e^{\frac{1}{2}\sum_{ij} A_{ij}s_i s_j} = \left\{\frac{1}{|\mathsf{A}|(2\pi)^N}\right\}^{1/2} \int \cdots \int e^{-\frac{1}{2}\sum_{ij}(\mathsf{A}^{-1})_{ij}\phi_i\phi_j + \sum_i \phi_i s_i} \left(\prod_{i=1}^{N} d\phi_i\right), \tag{2.2.5}$$

where $|\mathsf{A}|$ is the determinant of A. This identity is used to represent Ξ as

$$\begin{aligned}\mathscr{Z}_N &= \sum_{\substack{\{i_k\}\\ \{q_k=\pm e\}}} \frac{1}{N!} \exp\left[-\frac{\beta}{2} \sum_{\substack{j,k\\ 1\leq j,k\leq N}}{}' q_j U(i_j, i_k) q_k\right] \\ &= \left\{\frac{1}{|\mathsf{U}|(2\pi)^N}\right\}^{1/2} \sum_{\substack{\{i_k\}\\ \{q_k=\pm e\}}} \frac{1}{N!} \int \cdots \int e^{-\frac{1}{2}\beta^{-1}\sum_{jk}\phi_{i_j}(\mathsf{U}^{-1})_{jk}\phi_{i_k} + i\sum_k \phi_{i_k} q_k} \left(\prod_{k=1}^{N} d\phi_{i_k}\right),\end{aligned} \tag{2.2.6}$$

in which U is the matrix with entries given in Eq. (2.2.1) and the factor of i on the right-hand side is necessary for consistency of signs in Eqs. (2.2.5) and (2.2.6).

2.2.2 *Continuum Limit*

Consider first the determination of U^{-1}. In the case of continuous variables, we have the definition [18]

$$\int \mathsf{U}^{-1}(\mathbf{x}, \mathbf{y})\mathsf{U}(\mathbf{y}, \mathbf{z})\, d\mathbf{y} = \delta(\mathbf{x} - \mathbf{z}), \tag{2.2.7}$$

in which $\delta(x)$ is the Dirac delta-function. Since, according to Eq. (2.2.1),

$$U(\mathbf{x}, \mathbf{y}) = \frac{1}{4\pi|\mathbf{x} - \mathbf{y}|}, \tag{2.2.8}$$

Equation (2.2.7) is seen to be the definition of the fundamental solution for Poisson's equation:

$$\nabla^2 U(\mathbf{x} - \mathbf{x}') = -\delta(\mathbf{x} - \mathbf{x}'). \tag{2.2.9}$$

Hence,

$$\mathsf{U}^{-1}(\mathbf{x}, \mathbf{y}) = -\delta(\mathbf{x} - \mathbf{y})\nabla^2_{\mathbf{x}}. \tag{2.2.10}$$

Thus, in the continuum limit, the first term in the exponential on the right-hand side of Eq. (2.2.6) becomes

$$\frac{1}{2\beta}\sum_{jk}\phi_{i_j}(\mathsf{U}^{-1})_{jk}\phi_{i_k} \to \frac{1}{2\beta}\iint \phi(\mathbf{x})\,\mathsf{U}^{-1}(\mathbf{x},\mathbf{y})\,\phi(\mathbf{y})\,d\mathbf{x}\,d\mathbf{y}$$

$$= -\frac{1}{2\beta}\int \phi(\mathbf{x})\nabla^2\phi(\mathbf{x})\,d\mathbf{x}. \qquad (2.2.11)$$

The summation over the q_k on the right-hand side can now be carried out:

$$\sum_{\substack{\{i_k\}\\ \{q_k=\pm e\}}} \exp\left(i\sum_{k=1}^{N}\phi_{i_k}q_k\right) = \sum_{\substack{\{i_k\}\\ \{q_k=\pm e\}}}\prod_{k=1}^{N} e^{i\phi_{i_k}q_k} = \sum_{\{i_k\}}\prod_{k=1}^{N}\left(e^{-ie\phi_{i_k}} + e^{ie\phi_{i_k}}\right)$$

$$= \sum_{\{i_k\}}\prod_{k=1}^{N}\left[2\cos(e\phi_{i_k})\right]$$

$$= \left[\sum_{i_k} 2\cos(e\phi_{i_k})\right]^N. \qquad (2.2.12)$$

In the continuum limit, the summation becomes an integral and this term simplifies to

$$\sum_{\substack{\{i_k\}\\ \{q_k=\pm e\}}} \exp\left(i\sum_{k=1}^{N}\phi_{i_k}q_k\right) = \left\{\frac{1}{a^d}\int 2\cos\left[e\phi(\mathbf{x})\right]d\mathbf{x}\right\}^N. \qquad (2.2.13)$$

Carrying out the summation over N in Eq. (2.2.3) yields the continuum limit of the grand canonical partition function Ξ:

$$\Xi = \left\{\frac{1}{|\mathsf{U}|(2\pi)^N}\right\}^{1/2}\int \exp\left[\frac{1}{2\beta}\int \phi\nabla^2\phi\,d\mathbf{x}\right]\sum_{N=0}^{\infty}\frac{z^N a^{Nd}}{N!}\left\{\frac{1}{a^d}\int 2\cos\left[e\phi(\mathbf{x})\right]d\mathbf{x}\right\}^N D[\phi]$$

$$= \left\{\frac{1}{|\mathsf{U}|(2\pi)^N}\right\}^{1/2}\int \exp\left[\frac{1}{2\beta}\int \phi\nabla^2\phi\,d\mathbf{x}\right]\exp\left[z\int 2\cos(e\phi)\,d\mathbf{x}\right]D[\phi]$$

$$= \left\{\frac{1}{|\mathsf{U}|(2\pi)^N}\right\}^{1/2}\int \exp\left\{\int\left[\frac{1}{2\beta}\phi\nabla^2\phi + 2z\cos(e\phi)\right]d\mathbf{x}\right\}D[\phi]. \qquad (2.2.14)$$

The right-hand side of this equation is the functional integral for the Debye problem. The first term represents a ‘kinetic energy’, which results from the Coulomb interaction, and the second term a ‘potential energy’, which arises from the Hubbard–Stratonovich transformation. The prefactor is not important for calculating thermodynamic averages, and will be omitted in what follows. A somewhat more concise form of this representation is obtained by introducing a new field variable $\phi = \varphi\sqrt{\beta}$, in which case we obtain

$$\Xi = \int \exp\left\{\int \left[\tfrac{1}{2}\varphi\nabla^2\varphi + 2z\cos(\alpha\varphi)\right] d\mathbf{x}\right\} D[\varphi], \tag{2.2.15}$$

with $\alpha = \mathrm{e}/\sqrt{k_B T}$.

There are several advantages to the functional integral representation of a statistical mechanical problem over other formulations:

1. A large class of problems can be represented as functional integrals, ranging from equilibrium statistical mechanics, as the Debye problem, to non-equilibrium statistical dynamics. The Ising model serves as the canonical example for this approach, and the expression obtained in Eq. (2.2.15) has several formal similarities with the functional integral for the Ising model.
2. Mean-field limits are straightforward to identify, for example, as Gaussian field theories, corrections to which can be evaluated with various expansions. For the case at hand, the mean-field limit corresponds to the Debye–Hückel theory. This will shown explicitly in the next section.
3. Renormalization-group calculations can be carried out using either the Wilson or field-theoretic formalism. Time-dependence, vector fields, and other degrees of freedom enter such calculations simply as summations/integrals in the evaluation of individual terms. The structure of the RG expansion is determined by the polynomial terms in the functional integral.
4. An alternative form of the functional integral in Eq. (2.2.15) is obtained by performing an integration by parts on the first term:

$$\Xi = \int \exp\left\{-\int \left[\tfrac{1}{2}(\nabla\varphi)^2 - 2z\cos(\alpha\varphi)\right] d\mathbf{x}\right\} D[\varphi]. \tag{2.2.16}$$

In writing this expression, we have neglected the surface term, which is a finite constant and so does not affect our subsequent calculation.

2.3 Mean-Field Approximation

The simplest evaluation of the functional integral (2.2.15) is to expand the cosine function and retain terms only to quadratic order:

$$\cos(\alpha\varphi) = 1 - \frac{(\alpha\varphi)^2}{2} + \cdots, \tag{2.3.1}$$

which yields a Gaussian field theory,

$$\begin{aligned}\Xi &= \int \exp\left\{\frac{1}{2}\int \left[\varphi\nabla^2\varphi + 4z - 2z(\alpha\varphi)^2\right] d\mathbf{x}\right\} D[\varphi] \\ &= \mathrm{e}^{2zV} \int \exp\left[\frac{1}{2}\int \left(\varphi\nabla^2\varphi - \xi^{-2}\varphi^2\right) d\mathbf{x}\right] D[\varphi],\end{aligned} \tag{2.3.2}$$

where V is the volume of the system and $\xi^{-2} = 2z\alpha^2$. We will again neglect the constant prefactor, since we are interested only in averages.

The functional Gaussian integrals in Eq. (2.3.2) are carried out by transforming to a Fourier representation of decoupled modes. For a finite volume V, the Fourier transform $\varphi_{\mathbf{k}}$ of $\varphi(\mathbf{x})$ is

$$\varphi_{\mathbf{k}} = \int_V \varphi(\mathbf{x}) e^{-i\mathbf{k}\cdot\mathbf{x}}\, d\mathbf{x}, \tag{2.3.3}$$

where, since $\varphi(\mathbf{x})$ is real, we have that $\varphi_{-\mathbf{k}} = \varphi^*_{\mathbf{k}}$. The inverse Fourier transform is

$$\varphi(\mathbf{x}) = \frac{1}{V}\sum_{\mathbf{k}} \varphi_{\mathbf{k}} e^{i\mathbf{k}\cdot\mathbf{x}}. \tag{2.3.4}$$

The largest wavevector in this summation is $\Lambda \equiv 2\pi/a$, where a is the lattice constant on our Cartesian lattice, and the smallest is $2\pi/L$, where $V = L^3$ is the volume of the system and which approaches zero as $L \to \infty$. These are referred to as *ultraviolet* and *infrared* cutoffs, respectively. Critical quantities should not depend on the values of these cutoffs. Since there is one wavevector per volume $(2\pi/L)^d$ in $\mathbf{k}$-space, summations over $\mathbf{k}$ are converted into integrals according to

$$\sum_{\mathbf{k}} = \int d\mathbf{k}\left(\frac{L}{2\pi}\right)^d = V\int \frac{d\mathbf{k}}{(2\pi)^d}. \tag{2.3.5}$$

This transcription is exact only in the thermodynamic limit ($V \to \infty$). The transformed grand canonical partition function thereby reads

$$\Xi = \int \prod_{\mathbf{k}} d\varphi_{\mathbf{k}} \exp\left[-\frac{1}{2V}\sum_{\mathbf{k}}(k^2+\xi^{-2})|\varphi_{\mathbf{k}}|^2\right]. \tag{2.3.6}$$

Consider now the evaluation of the two-point correlation function $\langle\varphi(\mathbf{q})\varphi(\mathbf{q}')\rangle$ which, in the limit $V \to \infty$, is calculated as

$$\begin{aligned}\langle\varphi(\mathbf{q})\varphi(\mathbf{q}')\rangle &= \frac{1}{\Xi}\int D\varphi(\mathbf{k})\left[\varphi(\mathbf{q})\varphi(\mathbf{q}')\right]\exp\left[-\int_0^\Lambda \frac{d\mathbf{k}}{(2\pi)^d}(k^2+\xi^{-2})|\varphi(\mathbf{k})|^2\right]\\ &= \frac{\delta(\mathbf{q}+\mathbf{q}')}{q^2+\xi^{-2}}.\end{aligned} \tag{2.3.7}$$

Performing the Fourier transform yields the real-space correlation function

$$\langle\varphi(\mathbf{r})\varphi(\mathbf{0})\rangle = \int \frac{d\mathbf{q}}{(2\pi)^3}\frac{e^{-i\mathbf{q}\cdot\mathbf{r}}}{q^2+\xi^{-2}} = \frac{e^{-r/\xi}}{4\pi r}. \tag{2.3.8}$$

The quantity

$$\xi^{-2} = 2z\alpha^2 = \frac{2ze^2}{k_B T} \tag{2.3.9}$$

is seen to correspond to a screening length. To obtain an explicit expression for this quantity, we need an expression for the fugacity z. This can be obtained from the following two standard statistical mechanical relations for the grand canonical partition function

$$\langle N\rangle = \frac{1}{\Xi}\sum_{N=0}^{\infty} N z^N a^{Nd} \mathscr{Z}_N = z\frac{\partial \ln \Xi}{\partial z}, \tag{2.3.10}$$

$$\frac{PV}{k_B T} = \ln \Xi = 2zV + \ln\left\{\int \exp\left[\frac{1}{2}\int\left(\varphi\nabla^2\varphi - \xi^{-2}\varphi^2\right)d\mathbf{x}\right]D[\varphi]\right\}, \tag{2.3.11}$$

where P is the pressure of the system and we have used Eq. (2.3.2) in the second equation. We now consider the high temperature limit, in which case ξ^{-2} is small. Since the second term on the right-hand side of Eq. (2.3.11) is the only term with an explicit z-dependence, this limit suppresses the contribution from this term in the derivative in Eq. (2.3.10), leaving

$$\langle N\rangle = 2zV, \tag{2.3.12}$$

or

$$2z = \frac{\langle N\rangle}{V} = n_0, \tag{2.3.13}$$

where n_0 is the average particle density. Hence, the decay length reduces to

$$\xi = \left(\frac{k_B T}{n_0 e^2}\right)^{1/2}, \tag{2.3.14}$$

which is the standard result for the Debye length [6]. The validity of this results rests on the assumption that the $\cos(\alpha\varphi)$ is a slow varying function of position, which is certainly true at high temperatures.

References

1. J.B. Kogut, An introduction to lattice gauge theory and spin systems. Rev. Mod. Phys. **51**, 4 (1979)
2. J.M. Kosterlitz, D.J. Thouless, Ordering, metastability and phase transitions in two-dimensional systems. J. Phys. C: Solid State Phys. **6**, 1181 (1973)
3. P. Minnhagen, The two-dimensional Coulomb gas, vortex unbinding, and superfluid-superconducting films. Rev. Mod. Phys. **59**, 1001–1066 (1987)
4. S.T. Chui, J.D. Weeks, Phase transition in the two-dimensional Coulomb gas, and the interfacial roughening transition. Phys. Rev. B **14**, 4978 (1976)
5. D.J. Bishop, J.D. Reppy, Study of the superfluid transition in two-dimensional 4He films. Phys. Rev. Lett. **40**, 1727 (1978)
6. P. Debye, E. Hückel, Phys. Z. Sowjetunion **24**, 185 (1923)
7. J.P. Verboncoeur, Particle simulation of plasmas: review and advances. Plasma Phys. Controlled Fusion **47**, A231 (2005)
8. J. Fröhlich, Y.M. Park, Correlation inequalities and the thermodynamic limit for classical and quantum continuous systems. Comm. Math. Phys. **59**, 235–266, (1978)

9. J. Fröhlich, T. Spencer, The Kosterlitz-Thouless transition in two-dimensional abelian spin systems and the Coulomb gas. Comm. Math. Phys. **81**, 527–602, (1981)
10. W.S. Yang, Debye screening for two-dimensional Coulomb systems at high temperatures. J. Stat. Phys. **49**, 1–32 (1987)
11. F. Nicolò, P. Perfetti, The sine-Gordon field theory model at $\alpha^2 = 8\pi$, the nonsuperrenormalizable theory. Comm. Math. Phys. **123**, 425–452, (1989)
12. D.J. Amit, Y.Y. Goldschmidt, G. Grinstein, Renormalisation group analysis of the phase transition in 2D Coulomb gas, sine-Gordon theory and X–Y model. J. Phys. A: Math. Gen. **13** (1980)
13. J.M. Kosterlitz, The d-dimensional Coulomb gas and the roughening transition. J. Phys. C: Solid State Phys. **10**, 3753 (1977)
14. F.J. Wegner, A. Houghton, Renormalization group equation for critical phenomena. Phys. Rev. A **8**, 401 (1973)
15. I. Nándori, U.D. Jentschura, K. Sailer, G. Soff, Renormalization-group analysis of the generalized sine-Gordon model and of the Coulomb gas for $d>\sim 3$ dimensions. Phys. Rev. D **69**, (2003)
16. A.L. Kholodelenko, A.L. Beyerlein, Theory of symmetric electrolyte solutions: field-theoretic approach. Phys. Rev. A **34**, 3309 (1986)
17. T.S. Chang, D.D. Vvedensky, J.F. Nicoll, Differential renormalization-group generators for static and dynamic critical phenomena. Phys. Rep. **217**, 279 (1992)
18. S.F. Edwards, The statistical thermodynamics of a gas with long and short-range forces. Phil. Mag. **46**, 1171 (1959)

Chapter 3
Renormalization Group Analysis

3.1 Coarse-Graining of the Partition Function

The renormalization-group (RG) transformation of the functional integral in Eq. (2.2.15) will be performed by carrying out an infinitesimal form of a loop expansion, which subsumes irreducible terms in the usual diagrammatic representation [1]. With the grand canonical partition function in Eqs. (2.2.15) and (2.2.16) written in the form

$$\Xi = \int \exp\Big[-A\big(\{\varphi(\mathbf{k})\}\big)\Big]\, D[\varphi(\mathbf{k})], \tag{3.1.1}$$

where

$$A = \frac{1}{2}\int \frac{d\mathbf{k}}{(2\pi)^d} k^2 \varphi(\mathbf{k})\varphi(-\mathbf{k}) - 2z \int \cos\left[\alpha \int \frac{d\mathbf{k}}{(2\pi)^d} \mathrm{e}^{\mathrm{i}\mathbf{k}\cdot\mathbf{x}} \varphi(\mathbf{k})\right] d\mathbf{x}, \tag{3.1.2}$$

the coarse graining part of the RG transformation within this formulation can be expressed in closed form as a differential equation with respect to the renormalization parameter ℓ,

$$\frac{\partial A}{\partial \ell} = \frac{1}{2}\int \frac{d\Omega}{(2\pi)^d} \ln\left[A_{\mathbf{q},-\mathbf{q}} - \int \frac{d\mathbf{p}}{(2\pi)^d} \int \frac{d\mathbf{p}'}{(2\pi)^d} A_{\mathbf{q},\mathbf{p}}\, A^{-1}_{\mathbf{p},\mathbf{p}'}\, A_{\mathbf{p}',-\mathbf{q}}\right], \tag{3.1.3}$$

in which

$$A_{\mathbf{k},\mathbf{k}'} \equiv \frac{\delta^2 A}{\delta\varphi(\mathbf{k})\delta\varphi(\mathbf{k}')} \tag{3.1.4}$$

is the second functional derivative of A with respect to the Fourier components of the fluctuating field, the integrals of $\mathbf{p}$ and $\mathbf{p}'$ are over the range $1 \leq p, p' \leq \Lambda$, the integral of Ω is over the unit sphere in d-dimensions which, in our case, is $d = 3$. The second functional derivatives in Eq. (3.1.4) are

E. Barkhudarov, *Renormalization Group Analysis of Equilibrium and Non-equilibrium Charged Systems*, Springer Theses,
DOI: 10.1007/978-3-319-06154-2_3,

$$\frac{\delta^2 A}{\delta\varphi(\mathbf{k}'')\delta\varphi(\mathbf{k}')} = k'^2\delta(\mathbf{k}''+\mathbf{k}') + f(\mathbf{k}''+\mathbf{k}'), \tag{3.1.5}$$

where we have introduced the notation

$$f(\mathbf{k}''+\mathbf{k}') \equiv 2z\alpha^2 \int \cos\left[\alpha \int \frac{d\mathbf{k}}{(2\pi)^d} \mathrm{e}^{\mathrm{i}\mathbf{k}\cdot\mathbf{x}} \varphi(\mathbf{k})\right] \mathrm{e}^{\mathrm{i}(\mathbf{k}''+\mathbf{k}')\cdot\mathbf{x}}\, d\mathbf{x}\,. \tag{3.1.6}$$

The functional derivatives in Eq. (3.1.3) are therefore given by

$$A_{\mathbf{q},-\mathbf{q}} = 1 + f(\mathbf{0}), \tag{3.1.7}$$
$$A_{\mathbf{q},\mathbf{p}} = f(\mathbf{q}+\mathbf{p}), \tag{3.1.8}$$
$$A_{\mathbf{p},\mathbf{p}'} = p^2\delta(\mathbf{p}+\mathbf{p}') + f(\mathbf{p}+\mathbf{p}'), \tag{3.1.9}$$
$$A_{\mathbf{p},-\mathbf{q}} = f(\mathbf{p}-\mathbf{q})\,. \tag{3.1.10}$$

The remaining required quantities, $A^{-1}_{\mathbf{p},\mathbf{p}'}$, are the elements of the matrix inverse of the coefficients in Eq. (3.1.9). Since we will carry out the coarse-graining transformation only to order f^2, which is already accounted for by $A_{\mathbf{q},\mathbf{p}}$ and $A_{\mathbf{p},-\mathbf{q}}$, we require only the leading term in this quantity, which is

$$A^{-1}_{\mathbf{p},\mathbf{p}'} = \frac{1}{p^2}\,\delta(\mathbf{p}+\mathbf{p}') + \cdots\,. \tag{3.1.11}$$

Substituting the expressions in Eqs. (3.1.7), (3.1.8), (3.1.10), and (3.1.11) into Eq. (3.1.3) gives

$$\begin{aligned}
&\int \frac{d\Omega}{(2\pi)^d} \ln\left[1 + f(\mathbf{0}) - \int \frac{d\mathbf{p}}{(2\pi)^d} \int \frac{d\mathbf{p}'}{(2\pi)^d} f(\mathbf{q}+\mathbf{p}) \frac{1}{p^2} \delta(\mathbf{p}+\mathbf{p}') f(\mathbf{p}'-\mathbf{q})\right] \\
&\quad = \int \frac{d\Omega}{(2\pi)^d} \ln\left[1 + f(\mathbf{0}) - \int \frac{d\mathbf{p}}{(2\pi)^d} \frac{f(\mathbf{q}+\mathbf{p})f(-\mathbf{p}-\mathbf{q})}{p^2}\right] \\
&\quad = \int \frac{d\Omega}{(2\pi)^d} \left[f(\mathbf{0}) - \tfrac{1}{2}f^2(\mathbf{0}) - \int \frac{d\mathbf{p}}{(2\pi)^d} \frac{f(\mathbf{q}+\mathbf{p})f(-\mathbf{p}-\mathbf{q})}{p^2} + \cdots\right] \\
&\quad = \left[f(\mathbf{0}) - \tfrac{1}{2}f^2(\mathbf{0})\right]K_d - \int \frac{d\Omega}{(2\pi)^d} \int \frac{d\mathbf{p}}{(2\pi)^d} \frac{f(\mathbf{q}+\mathbf{p})f(-\mathbf{p}-\mathbf{q})}{p^2} + \cdots\,,
\end{aligned} \tag{3.1.12}$$

in which $K_d = S_d/(2\pi)^d$ and S_d is the surface area of the unit sphere in d dimensions. The term $f(\mathbf{0})$, according to Eq. (3.1.6), is given by

$$f(\mathbf{0}) = 2z\alpha^2 \int \cos\bigl[\alpha\varphi(\mathbf{x})\bigr]\, d\mathbf{x}\,. \tag{3.1.13}$$

Similarly, $f^2(\mathbf{0})$ is

$$f^2(\mathbf{0}) = 4z^2\alpha^4 \int \cos\big[\alpha\varphi(\mathbf{x})\big]\,d\mathbf{x} \int \cos\big[\alpha\varphi(\mathbf{x}')\big]\,d\mathbf{x}'$$
$$= 4z^2\alpha^4 \int d\mathbf{x} \int d\mathbf{x}' \cos\big[\alpha\varphi(\mathbf{x})\big]\cos\big[\alpha\varphi(\mathbf{x}')\big]. \qquad (3.1.14)$$

By invoking the trigonometric identity

$$\cos x \cos y = \tfrac{1}{2}\big[\cos(x+y) + \cos(x-y)\big], \qquad (3.1.15)$$

this integral can be written as

$$f^2(\mathbf{0}) = 2z^2\alpha^4 \int d\mathbf{x} \int d\mathbf{x}' \cos\{\alpha\big[\varphi(\mathbf{x}) + \varphi(\mathbf{x}')\big]\}$$
$$+ 2z^2\alpha^4 \int d\mathbf{x} \int d\mathbf{x}' \cos\{\alpha\big[\varphi(\mathbf{x}) - \varphi(\mathbf{x}')\big]\} \qquad (3.1.16)$$

The approximate evaluation of the integrals on the right-hand side of this equation can be carried out by introducing relative and average coordinates [2, 3],

$$\boldsymbol{\xi} = \mathbf{x} - \mathbf{x}', \quad \boldsymbol{\eta} = \tfrac{1}{2}(\mathbf{x} + \mathbf{x}'), \qquad (3.1.17)$$

in terms of which we have

$$\varphi(\mathbf{x}) = \varphi(\boldsymbol{\eta} + \tfrac{1}{2}\boldsymbol{\xi}), \quad \varphi(\mathbf{x}') = \varphi(\boldsymbol{\eta} - \tfrac{1}{2}\boldsymbol{\xi}). \qquad (3.1.18)$$

If we now use the fact that after several RG transformations φ becomes a slowly-varying function of its argument,

$$\varphi(\boldsymbol{\eta} \pm \tfrac{1}{2}\boldsymbol{\xi}) \approx \varphi(\boldsymbol{\eta}) \pm \tfrac{1}{2}\boldsymbol{\xi}\cdot\nabla\varphi(\boldsymbol{\eta}), \qquad (3.1.19)$$

from which we obtain

$$\varphi(\mathbf{x}) + \varphi(\mathbf{x}') = \varphi(\boldsymbol{\eta} + \tfrac{1}{2}\boldsymbol{\xi}) + \varphi(\boldsymbol{\eta} - \tfrac{1}{2}\boldsymbol{\xi}) \approx 2\varphi(\boldsymbol{\eta}), \qquad (3.1.20)$$
$$\varphi(\mathbf{x}) - \varphi(\mathbf{x}') = \varphi(\boldsymbol{\eta} + \tfrac{1}{2}\boldsymbol{\xi}) - \varphi(\boldsymbol{\eta} - \tfrac{1}{2}\boldsymbol{\xi}) \approx \boldsymbol{\xi}\cdot\nabla\varphi(\boldsymbol{\eta}). \qquad (3.1.21)$$

By using these results and the transformation in Eq. (3.1.17), we obtain

$$f^2(\mathbf{0}) = 2z^2\alpha^4 \int d\boldsymbol{\xi} \int \cos\big[2\alpha\varphi(\boldsymbol{\eta})\big]\,d\boldsymbol{\eta} + 2z^2\alpha^4 \int d\boldsymbol{\eta} \int \cos\big[\alpha\boldsymbol{\xi}\cdot\nabla\varphi(\boldsymbol{\eta})\big]\,d\boldsymbol{\xi} \qquad (3.1.22)$$

The integral over $\boldsymbol{\xi}$ in the first term on the right-hand side of this equation is the volume V of the system. Hence,

$$2z^2\alpha^4 \int d\boldsymbol{\xi} \int \cos\big[2\alpha\varphi(\boldsymbol{\eta})\big]\,d\boldsymbol{\eta} = 2z^2\alpha^4 V \int \cos\big[2\alpha\varphi(\boldsymbol{\eta})\big]\,d\boldsymbol{\eta}. \qquad (3.1.23)$$

Then, by expanding the cosine function in the second integral and retaining only the first two terms, we obtain

$$2z^2\alpha^4\int d\boldsymbol{\eta}\int d\boldsymbol{\xi}\Big\{1-\tfrac{1}{2}\big[\alpha\boldsymbol{\xi}\cdot\nabla\varphi(\boldsymbol{\eta})\big]^2\Big\}=2z^2\alpha^4V^2-z^2\alpha^6I_1\int\nabla\varphi(\boldsymbol{\eta})\cdot\nabla\varphi(\boldsymbol{\eta})\,d\boldsymbol{\eta},\tag{3.1.24}$$

in which we have defined

$$I_1=\int\xi^2\,d\boldsymbol{\xi}\,.\tag{3.1.25}$$

Thus,

$$f^2(\mathbf{0})=2z^2\alpha^4V^2+2z^2\alpha^4V\int\cos\big[2\alpha\varphi(\boldsymbol{\eta})\big]\,d\boldsymbol{\eta}-z^2\alpha^6I_1\int\nabla\varphi(\boldsymbol{\eta})\cdot\nabla\varphi(\boldsymbol{\eta})\,d\boldsymbol{\eta}\,.\tag{3.1.26}$$

The evaluation of the remaining integral on the right-hand side of Eq. (3.1.12) proceeds by writing

$$\begin{aligned}&\int\frac{d\Omega}{(2\pi)^d}\int\frac{d\mathbf{p}}{(2\pi)^d}\frac{f(\mathbf{q}+\mathbf{p})f(-\mathbf{p}-\mathbf{q})}{p^2}\\&=4\alpha^4z^2\int\frac{d\Omega}{(2\pi)^d}\int\frac{d\mathbf{p}}{(2\pi)^d}\frac{1}{p^2}\Big\{\int\cos\big[\alpha\varphi(\mathbf{x})\big]e^{i(\mathbf{q}+\mathbf{p})\cdot\mathbf{x}}\,d\mathbf{x}\Big\}\Big\{\int\cos\big[\alpha\varphi(\mathbf{x}')\big]e^{-i(\mathbf{q}+\mathbf{p})\cdot\mathbf{x}'}\,d\mathbf{x}'\Big\}\\&=4\alpha^4z^2\int d\mathbf{x}\int d\mathbf{x}'\cos\big[\alpha\varphi(\mathbf{x})\big]\cos\big[\alpha\varphi(\mathbf{x}')\big]\int\frac{d\Omega}{(2\pi)^d}e^{i\mathbf{q}\cdot(\mathbf{x}-\mathbf{x}')}\int\frac{d\mathbf{p}}{(2\pi)^d}\frac{e^{i\mathbf{p}\cdot(\mathbf{x}-\mathbf{x}')}}{p^2}\end{aligned}\tag{3.1.27}$$

The integrals over $\mathbf{q}$ and $\mathbf{p}$ will be represented as

$$I_2(\mathbf{x}-\mathbf{x}')\equiv\int\frac{d\Omega}{(2\pi)^d}e^{i\mathbf{q}\cdot(\mathbf{x}-\mathbf{x}')},\tag{3.1.28}$$

$$I_3(\mathbf{x}-\mathbf{x}')\equiv\int\frac{d\mathbf{p}}{(2\pi)^d}\frac{e^{i\mathbf{p}\cdot(\mathbf{x}-\mathbf{x}')}}{p^2}\,.\tag{3.1.29}$$

Again invoking the trigonometric identity in Eq. (3.1.15) enables us to write

$$\begin{aligned}&\int d\mathbf{x}\int d\mathbf{x}'\cos\big[\alpha\varphi(\mathbf{x})\big]\cos\big[\alpha\varphi(\mathbf{x}')\big]I_2(\mathbf{x}-\mathbf{x}')I_3(\mathbf{x}-\mathbf{x}')\\&\quad=\frac{1}{2}\int d\mathbf{x}\int d\mathbf{x}'\cos\big\{\alpha\big[\varphi(\mathbf{x})+\varphi(\mathbf{x}')\big]\big\}I_2(\mathbf{x}-\mathbf{x}')I_3(\mathbf{x}-\mathbf{x}')\\&\quad\quad+\frac{1}{2}\int d\mathbf{x}\int d\mathbf{x}'\cos\big\{\alpha\big[\varphi(\mathbf{x})-\varphi(\mathbf{x}')\big]\big\}I_2(\mathbf{x}-\mathbf{x}')I_3(\mathbf{x}-\mathbf{x}')\,.\end{aligned}\tag{3.1.30}$$

Using these results and the transformation (3.1.17) in the integrals on the right-hand side of Eq. (3.1.30) then yields

$$\int d\boldsymbol{\eta}\int \cos\big[2\alpha\varphi(\boldsymbol{\eta})\big]I_2(\boldsymbol{\xi})I_3(\boldsymbol{\xi})\,d\boldsymbol{\xi}+\int d\boldsymbol{\eta}\int \cos\big[\alpha\boldsymbol{\xi}\cdot\nabla\varphi(\boldsymbol{\eta})\big]I_2(\boldsymbol{\xi})I_3(\boldsymbol{\xi})\,d\boldsymbol{\xi} \tag{3.1.31}$$

The first term may be written as

$$\left\{\int I_2(\boldsymbol{\xi})I_3(\boldsymbol{\xi})\,d\boldsymbol{\xi}\right\}\int \cos\big[2\alpha\varphi(\boldsymbol{\eta})\big]\,d\boldsymbol{\eta}\,. \tag{3.1.32}$$

Upon expansion of the cosine function in the second integral and retaining only the first two terms, we obtain

$$\begin{aligned}
&\int d\boldsymbol{\eta}\int d\boldsymbol{\xi}\left\{1-\tfrac{1}{2}\big[\alpha\boldsymbol{\xi}\cdot\nabla\varphi(\boldsymbol{\eta})\big]^2\right\}I_2(\boldsymbol{\xi})I_3(\boldsymbol{\xi})\\
&=\int d\boldsymbol{\eta}\int I_2(\boldsymbol{\xi})I_3(\boldsymbol{\xi})\,d\boldsymbol{\xi}-\frac{1}{2}\int d\boldsymbol{\eta}\int\big[\alpha\boldsymbol{\xi}\cdot\nabla\varphi(\boldsymbol{\eta})\big]^2 I_2(\boldsymbol{\xi})I_3(\boldsymbol{\xi})\,d\boldsymbol{\xi}\\
&=V\int I_2(\boldsymbol{\xi})I_3(\boldsymbol{\xi})\,d\boldsymbol{\xi}-\frac{\alpha^2}{2}\left\{\int \xi^2 I_2(\boldsymbol{\xi})I_3(\boldsymbol{\xi})\,d\boldsymbol{\xi}\right\}\int \nabla\varphi(\boldsymbol{\eta})\cdot\nabla\varphi(\boldsymbol{\eta})\,d\boldsymbol{\eta}\,.
\end{aligned}\tag{3.1.33}$$

Thus,

$$\begin{aligned}
&\int\frac{d\Omega}{(2\pi)^d}\int\frac{d\mathbf{p}}{(2\pi)^d}\frac{f(\mathbf{q}+\mathbf{p})f(-\mathbf{p}-\mathbf{q})}{p^2}\\
&=2\alpha^4z^2AV-\alpha^6z^2B\int\big[\nabla\varphi(\boldsymbol{\eta})\big]^2\,d\boldsymbol{\eta}+2\alpha^4z^2A\int\cos\big[2\alpha\varphi(\boldsymbol{\eta})\big]\,d\boldsymbol{\eta},
\end{aligned}\tag{3.1.34}$$

where we have defined

$$A\equiv\int I_2(\boldsymbol{\xi})I_3(\boldsymbol{\xi})\,d\boldsymbol{\xi},\quad B\equiv\int \xi^2 I_2(\boldsymbol{\xi})I_3(\boldsymbol{\xi})\,d\boldsymbol{\xi}\,. \tag{3.1.35}$$

By collecting the results in Eqs. (3.1.13), (3.1.26), and (3.1.34) and substituting into Eq. (3.1.12) yields the coarse-graining transformation to one-loop order:

$$\begin{aligned}
\frac{\partial A}{\partial \ell}=&-\tfrac{1}{2}z^2\alpha^4V^2K_d-\alpha^4z^2AV+z\alpha^2K_d\int\cos\big[\alpha\varphi(\mathbf{x})\big]\,d\mathbf{x}\\
&+\tfrac{1}{4}z^2\alpha^6I_1K_d\int\big[\nabla\varphi(\boldsymbol{\eta})\big]^2\,d\boldsymbol{\eta}+\tfrac{1}{2}z^2\alpha^6B\int\big[\nabla\varphi(\boldsymbol{\eta})\big]^2\,d\boldsymbol{\eta}\\
&-\tfrac{1}{2}z^2\alpha^4VK_d\int\cos\big[2\alpha\varphi(\boldsymbol{\eta})\big]\,d\boldsymbol{\eta}-z^2\alpha^4A\int\cos\big[2\alpha\varphi(\boldsymbol{\eta})\big]\,d\boldsymbol{\eta}\,.
\end{aligned}\tag{3.1.36}$$

3.2 Differential Scale Transformations

The first two terms on the right-hand side of Eq. (3.1.36) are constants and represent the effect of the integrated modes on the grand canonical partition function. These terms are important for the direct calculation of Ξ [1], but will not be considered here. The remaining terms represent the effect of coarse graining on the terms in A in Eq. (3.1.1), with the last two terms representing corrections to the 'bare' form of A. Therefore, the lowest order effect of coarse graining is given by the following subset of terms on Eq. (3.1.36):

$$\frac{\partial A}{\partial \ell} = \tfrac{1}{2}z^2\alpha^6(\frac{I_1 K_d}{2} + B)\int \left[\nabla\varphi(\mathbf{x})\right]^2 d\mathbf{x} + z\alpha^2 K_d \int \cos\left[\alpha\varphi(\mathbf{x})\right] d\mathbf{x} \,. \quad (3.2.1)$$

This equation must be supplemented by two additional operations to obtain the complete recursion relations: the rescaling of $\mathbf{k}$ to restore the original ranges of integration, and the rescaling of φ to maintain the coefficient of the quadratic term in A. The latter operation is necessary to maintain the spatial range of the fluctuations under the RG transformation [cf. Eq. (2.3.2) et seq.]. The rescalings of the wave vector and the field are given by:

$$\mathbf{k}' = b\mathbf{k}, \quad (3.2.2)$$

$$\varphi(b^{-1}\mathbf{k}) = b^{(d+2)/2}\varphi'(\mathbf{k}) \,, \quad (3.2.3)$$

where $b = e^{\delta\ell}$. Denoting the rescaling operator by $\mathscr{R}$, the change in A induced by the scale changes is defined as

$$A\left(\{\varphi'(\mathbf{k}')\}\right) + \Delta A\left(\{\varphi'(\mathbf{k}')\}\right) = \mathscr{R}A\left(\{\varphi(\mathbf{k})\}\right) \,. \quad (3.2.4)$$

After applying the transformations set by Eqs. (3.2.2) and (3.2.3) we obtain:

$$\begin{aligned} \mathscr{R}A\left(\{\varphi(\mathbf{k})\}\right) = &\frac{1}{2}\int \frac{d\mathbf{k}'}{(2\pi)^d}\left(k'\right)^2 \varphi'(\mathbf{k}')\varphi'(-\mathbf{k}') \\ &- 2zb^d \int \cos\left[\alpha b^{1-d/2}\int \frac{d\mathbf{k}'}{(2\pi)^d} e^{i\mathbf{k}'\cdot\mathbf{x}'}\varphi'(\mathbf{k}')\right] d\mathbf{x}' \,. \end{aligned}$$

By considering an infinitesimal transformation and taking the limit of $\delta\ell \to 0$ yields the differential rescaling of A

$$\begin{aligned} \lim_{\delta\ell\to 0} \frac{\mathscr{R}A - A}{\delta\ell} = &- 2zd\int \cos\left[\alpha\varphi'(\mathbf{x}')\right] d\mathbf{x}' \\ &+ 2z\int \sin\left[\alpha\varphi'(\mathbf{x}')\right]\alpha\varphi'(\mathbf{x}')\left(1 - \frac{d}{2}\right) d\mathbf{x}' \,. \end{aligned}$$

Thus, the total change in A is

$$\frac{\partial A}{\partial \ell} = \frac{1}{2} z^2 \alpha^6 \left(\frac{I_1 K_d}{2} + B \right) \int \left[\nabla \varphi(\mathbf{x}) \right]^2 d\mathbf{x} - 2z \left(d - \frac{\alpha^2 K_d}{2} \right) \int \cos\left[\alpha \varphi(\mathbf{x}) \right] d\mathbf{x}$$
$$+ 2z \int \sin\left[\alpha \varphi(\mathbf{x}) \right] \alpha \varphi(\mathbf{x}) \left(1 - \frac{d}{2} \right) d\mathbf{x}\,.$$

The recursion relations can now be written:

$$\frac{d\varphi}{d\ell} = \frac{1}{2} z^2 \alpha^6 \left(\frac{I_1 K_d}{2} + B \right) \varphi, \tag{3.2.5}$$

$$\frac{dz}{d\ell} = z \left(d - \frac{\alpha^2 K_d}{2} \right), \tag{3.2.6}$$

$$\frac{d\,(\alpha\varphi)}{d\ell} = \alpha\varphi \left(1 - \frac{d}{2} \right). \tag{3.2.7}$$

By demanding the invariance of the kinetic term under RG transformation, we restrict the phase space to the flow in the coupling phase space only. The resultant recursion relations are

$$\frac{dz}{d\ell} = z \left(d - \frac{\alpha^2 K_d}{2} \right), \tag{3.2.8}$$

$$\frac{d\alpha}{d\ell} = \alpha \left(\left(1 - \frac{d}{2} \right) - \frac{1}{2} z^2 \alpha^6 \left(\frac{I_1 K_d}{2} + B \right) \right). \tag{3.2.9}$$

3.3 Recursion Relations and Fixed Points

In the differential equations derived in the previous section we noted that the cut-off dependence is incorporated into the limits of the integrals such as $I_3\,(\mathrm{x})$. This is due to the fact that we have considered the change in the Hamiltonian due to eliminating integrating the Fourier mode around $|\mathrm{k}| = 1$ shell. The modes in the shell $1 < |\mathrm{k}| < \Lambda$ were averaged out using a saddle point approximation.

3.3.1 $d = 1$

The differential form of the recursion relations is:

$$\frac{dz}{d\ell} = z \left(d - \frac{\alpha^2 K_d}{2} \right), \tag{3.3.1}$$

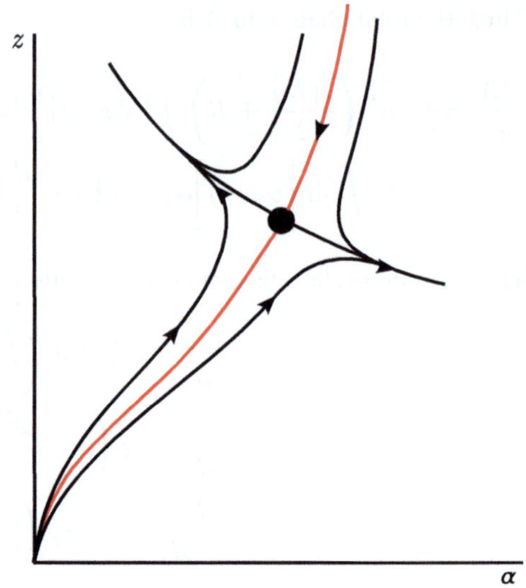

Fig. 3.1 Renormalization group trajectories for 1D Coulomb gas. The non-trivial fixed point $(z_+; \alpha_+)$, is a saddle point. There is only one trajectory which takes the couplings to this fixed point, depicted in *red*. Any perturbations away from this line will eventually drive the system away from that fixed point

$$\frac{d\alpha}{d\ell} = \alpha \left(\frac{1}{2} - \frac{z^2 \alpha^6}{8\pi} (I_1 + 2\pi B) \right). \tag{3.3.2}$$

By considering $\dot{z} = \dot{\alpha} = 0$, we can determine the critical points. Apart from the trivial fixed point $(z_0; \alpha_0) = (0; 0)$, there are four non-trivial points $(z_\pm; \alpha_\pm) = \left(\pm\pi^2 (I_1 + 2\pi B)^{-\frac{1}{2}} ; \pm\sqrt{2\pi} \right)$. However we are only interested in the flow in the positive quadrant of the phase space, so we only have to investigate the stability of two points: $(z_0; \alpha_0)$ and $(z_+; \alpha_+)$. By linearizing the equations around a fixed point, $z \to z_* + \varepsilon$ and $\alpha \to \alpha_* + \delta$, its stability is determined. Thus, the trivial point is characterized by the following linear differential equations:

$$\frac{d}{d\ell} \begin{pmatrix} \varepsilon \\ \delta \end{pmatrix} = \begin{pmatrix} 1 & 0 \\ 0 & \frac{1}{2} \end{pmatrix} \begin{pmatrix} \varepsilon \\ \delta \end{pmatrix}. \tag{3.3.3}$$

The matrix is diagonal therefore its eigenvectors and the corresponding eigenvalues, λ_1 and λ_2, can be determined immediately. Since $\lambda_1, \lambda_2 > 0$, we conclude that the point $(z_0; \alpha_0)$ is unstable. By linearizing around $(z_+; \alpha_+)$ the dynamics is governed by the following coupled differential equations:

$$\frac{d}{d\ell} \begin{pmatrix} \varepsilon \\ \delta \end{pmatrix} = \begin{pmatrix} 0 & -\frac{\alpha_+ z_+}{2} \\ -\frac{\alpha_+}{z_+} & -3 \end{pmatrix} \begin{pmatrix} \varepsilon \\ \delta \end{pmatrix}. \tag{3.3.4}$$

The associated eigenvalues are $\lambda_{1,2} = -\frac{3}{2} \pm \frac{1}{2}\sqrt{9 + 4\pi}$. We conclude that $(z_+; \alpha_+)$ is a saddle point. The phase space diagram associated with 1D Coulomb gas is depicted in Fig. 3.1.

This result qualitatively agrees with the previous study carried out by Kosterlitz [4], where the author used real-space renormalization group technique. The phase-space diagram suggests that in $d = 1$ the Coulomb gas undergoes a metal-insulator phase transition. If the initial parameters of the system are to the left of the red line then the system is driven towards the infinite termperature regime where charges are unbound. This forms a metallic state. The system defined to the right of the diagram is driven towards the low temperature regime where it behaves as an insulator.

3.3.2 $d = 2$

It is known that in 2D, the Coulomb gas undergoes the Kosterlitz–Thouless transition. We can use the equations derived in the previous section to reproduce the well known phase-diagram describing the Kosterlitz–Thouless transition. Consider the phase space trajectories near the critical line.

$$\frac{dz}{d\ell} = \frac{1}{4\pi}\left(8\pi - \alpha^2\right) z, \tag{3.3.5}$$

$$\frac{d\alpha}{d\ell} = -\tfrac{1}{2} z^2 \alpha^7 \left(\frac{I_1}{4\pi} + B\right). \tag{3.3.6}$$

Make the substitution:

$$y = -8\pi + \alpha^2, \tag{3.3.7}$$

which leads to the following set of equations:

$$\frac{dz}{d\ell} = -\frac{1}{4\pi} yz, \tag{3.3.8}$$

$$\frac{dy}{d\ell} = -z^2 (8\pi + y)^4 \left(\frac{I_1}{4\pi} + B\right). \tag{3.3.9}$$

Close to the transition temperature, $y \ll 1$, so terms involving y of order 1 should be retained. The resultant equations are:

$$\frac{dz^2}{d\ell} = -\frac{yz^2}{4\pi}, \tag{3.3.10}$$

$$\frac{dy^2}{d\ell} = -yz^2 (8\pi)^4 \left(\frac{I_1}{4\pi} + B\right). \tag{3.3.11}$$

These equations can now be written:

$$\frac{dz^2}{d\ell} = c\frac{dy^2}{d\ell}, \quad \text{where } c = \frac{1}{(8\pi)^4 (I_1 + 4\pi B)}, \tag{3.3.12}$$

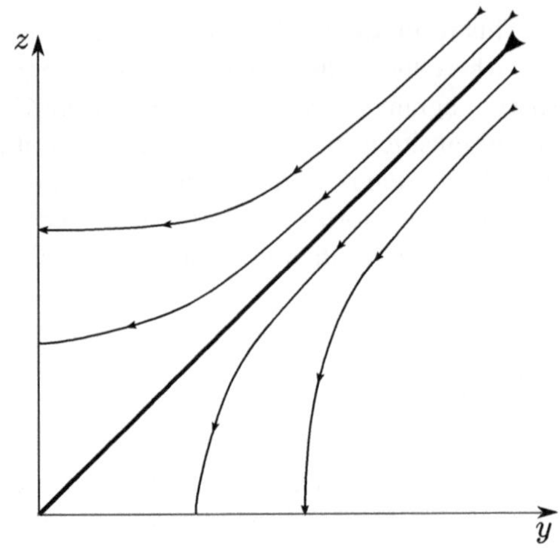

Fig. 3.2 Renormalization group trajectories for 2D Coulomb gas.The *thick line* of constant gradient is the critical line below which the system is in an insulating state. Above the critical line the system is in a metallic state and acts as a good conductor

which allows us to solve it. The solution is a hyperbola:

$$z^2 - cy^2 = const. \tag{3.3.13}$$

By rescaling the y variable the phase space flow is depicted in Fig. 3.2. The critical temperature is found along the trajectory $z^2 - cy^2 = 0$.

3.3.3 $d = 3$

It is known that in 3D Coulomb gas there is no phase transition. This happens because in the first term in the differential equation for α there is a sign change for $d \geq 3$. Thus, unless we are prepared to accept imaginary fugacity, there is no phase transition for dimensions above $d = 2$. This result agrees with the study carried out by Kosterlitz [4], where the author arrives at the same conclusion. The phase space flow is depicted in Fig. 3.3. The RG equation are:

$$\frac{dz}{d\ell} = \left(3 - \frac{\alpha^2}{4\pi^2}\right) z, \tag{3.3.14}$$

$$\frac{d\alpha}{d\ell} = -\frac{\alpha}{2}\left(1 + z^2\alpha^6\left(\frac{I_1}{4\pi^2} + B\right)\right). \tag{3.3.15}$$

There is no phase transition and trajectories are driven towards the high temperature regime. The system is in a metallic state throughout. Despite the fact that no transition takes place at $d \geq 3$, we can exploit renormalization group equations to study the behavior of other quantities such as the Debye length. We recall the definition, $\xi^{-2} = 2z\alpha^2$. Hence

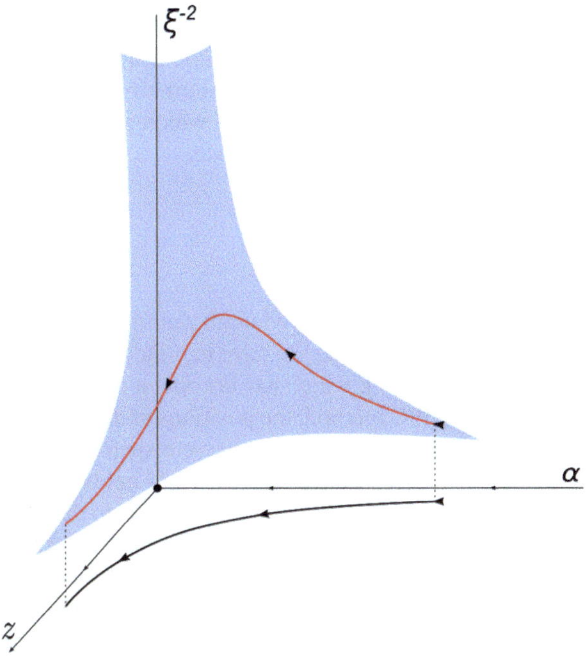

Fig. 3.3 The α–z plane illustrates the renormalization group trajectory for 3D Coulomb gas. The *blue surface* represents the Debye length as a function of two variables, $\xi^{-2}(z,\alpha) = 2z\alpha^2$. The behavior of the Debye length under RG is depicted by the *red curve* which lives on the $\xi^{-2}(z,\alpha)$ surface

$$\frac{d\xi^{-2}}{d\ell} = 2\alpha^2\left(d - \alpha^2\frac{K_d}{2}\right)z + 4z\alpha^2\left(\left(1 - \frac{d}{2}\right) - \frac{1}{2}z^2\alpha^6\left(\frac{I_1 K_d}{2} + B\right)\right). \tag{3.3.16}$$

The behavior of the Debye length is depicted by a red line in Fig. 3.3. Our renormalization group equations are valid for up to the second order in z and the definition of the Debye length is valid in the limit of the Debye–Hückel–Bjerrum theory (high temperature limit). So we choose a particular subspace of the phase-space where these two regimes hold. Coincidently, this regime corresponds to the phase-flow in the proximity of the trivial fixed point $(z_*, \alpha_*) = (0, 0)$. Thus, in $d = 3$, Eq. (3.3.16) gives a correct description of the behavior of the Debye length under coarse-graining:

$$\frac{d\xi}{d\ell} = \xi\left(\frac{\alpha^2 K_d}{2\pi^2} - 4\right) + \frac{\alpha^2}{2\xi^3}\left(\frac{I_1}{4\pi^2} + B\right). \tag{3.3.17}$$

The cut-off dependence is incorporated in the B integral. If we choose $\Lambda \to \infty$, the integral simplifies and becomes a function of the volume only. In order to study the behavior of the Debye length as a function of the UV cut-off we would require to modify our RG equation. Thus, in principle one could investigate variations of the Debye length as a function of the system size. Alternatively, one could investigate similar variations but with respect to the cut-off which is a linear scale associated with the lattice spacing of the original lattice model.

3.4 Conclusions and Remarks

The 3D result for the behavior of the Debye length can find an application of the RG analysis of the Coulomb gas to plasmas. We have in mind, in particular, the so-called particle-in-cell (PIC) method [5–7], which has become a powerful technique for the simulation of plasmas. In the PIC method, the positions and velocities of the particles are defined continuously, fields are defined at discrete spatial points, but both fields and particles are defined at discrete times. Particles and fields are advanced sequentially in time using fields interpolated from the discrete grid to the continuous particle locations. Source terms for the field equations are accumulated from the continuous particle locations to the discrete mesh locations. The fields are then advanced one time step, whereupon the loop repeats. The accuracy of the PIC method depends on the discretization in time and space. While the consequences of temporal discretization for a particular scheme, such as leap-frog method, are relatively straightforward to identify, the effect of spatial discretization is much more difficult to quantify [5]. Deviation from spatial continuity towards discretization induces noise into the the dynamics of the system. Moreover physical quantities which characterize plasma would also be expected to be affected by spacial discretization.

By drawing on the similarity of the Coulomb gas on a lattice and a specific realization of the PIC scheme, one can investigate the effect of changing the scales of the system on the physical parameters of the system. The lattice spacing and the system size enter naturally into the RG approach, as the the ultraviolet and infrared cut-offs, respectively. The RG equations for the Debye length with the effect of the length scales incorporated into the equations, which allows us to quantify how the coupling constants of the coarse-grained system are affected by particular choices of the system size and how this is reflected in the RG trajectories.

References

1. T.S. Chang, D.D. Vvedensky, J.F. Nicoll, Differential renormalization-group generators and dynamic critical phenomena. Phys. Rep. **217**, 279 (1992)
2. J.B. Kogut, An introduction to lattice gauge theory and spin systems. Rev. Mod. Phys. **51**(4), 659–713 (1979)
3. A.L. Kholodelenko, A.L. Beyerlein, Theory of symmetric electrolyte solutions: field-theoretic approach. Phys. Rev. A **34**, 3309 (1986)
4. J.M. Kosterlitz, The d-dimensional Coulomb gas and the roughening transition. J. Phys. C: Solid State Phys. **10**, 3753 (1977)

5. J.P. Verboncoeur, Particle simulation of plasmas: review and advances. Plasma Phys. Controlled Fusion **47**, A231 (2005)
6. D. Tskhakaya, The particle-in-cell methods, lecture notes. Lect. Notes Phy. **739**, 161–189 (2008)
7. C.K. Birdsall, Particle-in-cell charged-particle simulations, plus Monte Carlo collisions with neutral atoms, PIC-MCC. Trans. Plasma Sci. **19**(2), 65–85 (1991)

Part II
Non-Equilibrium Statistical Mechanics: Randomly Stirred Magnetohydrodynamics

Chapter 4
Turbulent Flows

4.1 Introduction

Turbulence is a widespread phenomenon in the dynamics of fluids. For decades, it has stimulated scientific research across various fields of mathematics, engineering and physics. Turbulent motion can be made visible by observing cloud formation in the atmosphere, for example. An overwhelming number of possible patterns in the motion underlies the complexity of this phenomenon. Turbulence is a state of a fluid which, in the case of neutral fluids, is governed by the Navier-Stokes equation. The problem of turbulence lies in the difficulty of describing a solution to the dynamical equation.

Turbulent motion can be characterized by its ability to stir fluids and dissipate kinetic energy. It is composed of eddies which are irregular zigzag patterns in the flow which can be visualized as a random swirling of the fluid, which typically arises when the speed of the flow exceeds a particular threshold. Understanding turbulent motion is of great importance in science. In engineering applications, for example in aerodynamics, a solid grasp of the phenomenon can improve the maneuverability of jet fighters or improve fuel efficiency for commercial airlines [1]. On the other hand, in internal combustion engines, turbulence enhances the fuel mixing with the oxidizers to augment efficiency [1]. There are even biological applications, which include studies of abrupt blood flows in heart's ventricles [1].

From the experience of our terrestrial world it is not apparent why one may want to study charged fluids, as oppose to the neutral fluids, since all electrical conductors are mainly solids. However, most of the universe is in the plasma state that can interact with magnetic fields which penetrate it [2]. Such systems often can be conveniently described by magnetohydrodynamics (MHD), which unifies the fields of electromagnetism and fluid mechanics. The scope of various phenomena which can be described by MHD is vast. Much research has gone into understanding solar systems, for example [3, 4]. In geophysics, MHD equations are applicable to conduction cores of the planets where the magnetic field is generated by dynamo action [5].

E. Barkhudarov, *Renormalization Group Analysis of Equilibrium and Non-equilibrium Charged Systems*, Springer Theses,
DOI: 10.1007/978-3-319-06154-2_4,

MHD turbulence is intimately related to hydrodynamic turbulence, in the sense that often the formalism developed for the former can be generalized to the latter. The study of turbulence can be taken along two different lines: one, which focuses on practical applications of turbulence and treats it as a technical problem. and the other, which treats turbulence more as a physical phenomenon and tries to develop tools towards its understanding. Over the past decades, with the development of computer power, numerical simulations have become an indispensable tool in turbulence research and often is the only way to check the validity of analytical models. However, numerically attainable Reynolds numbers are still too small to understand real turbulence and, thus, studies of high Reynolds number turbulence are of a more academic nature [6]. In fact, numerical simulations are of even greater importance to MHD turbulence, since there do not exist any practical laboratory set ups for such systems.

In this thesis we aim to study turbulence in charged systems using analytical tools. There exists a number of different approaches, each with its strengths and limitations. In order to appreciate the wealth of techniques which exist to date it is instructive to proceed towards a more quantitative description of the problem.

4.2 Mathematical Formulation

In order to facilitate a quantitative discussion of turbulence in neutral and charged flows, we begin with an introduction to the basic equations of motion. The dynamics of neutral fluids is governed by the Navier-Stokes (NS) equation [7, 8]:

$$\frac{\partial \mathbf{v}}{\partial t} + \mathbf{v} \cdot \nabla \mathbf{v} = -\frac{1}{\rho} \nabla p + \nu \nabla^2 \mathbf{v} + \frac{1}{\rho} \mathbf{F}, \tag{4.2.1}$$

where $\mathbf{v}$ is the velocity field, ρ is the density of the fluid, which can be a function of both position and time, ν is the kinematic viscosity, and $\mathbf{F}$ represents any additional external forces which act on the fluid. The specification of this force is subject to the problem at hand. The NS equation can be seen to express the conservation of momentum. Physical flows should also satisfy the continuity equation, which represents conservation of mass in the flow:

$$\frac{\partial \rho}{\partial t} + \nabla \cdot (\rho \mathbf{v}) = 0. \tag{4.2.2}$$

In our analysis we will be concerned primarily with incompressible flows, where ρ is a constant. Thus, the continuity equation translates to the condition of incompressibility of the flow:

$$\nabla \cdot \mathbf{v} = 0. \tag{4.2.3}$$

This means that the form of the velocity field is solenoidal, meaning that the divergence of the field $\mathbf{v}$ is zero at all points in space. Further specification of the problem must include boundary conditions. We are primarily concerned with unbounded flows where the appropriate boundary condition is:

$$\mathbf{v} \to 0; \quad \frac{\partial \mathbf{v}}{\partial x_i} \to 0 \ \ \forall i \quad \text{as} \quad \mathbf{x} \to \infty. \tag{4.2.4}$$

To appreciate the role of various terms in the NS equation it is convenient to re-write it in a dimensionless form. This can be achieved by the following transformations:

$$L^{-1}\mathbf{x} = \mathbf{x}', \qquad L^{-1}Ut = t', \qquad U^{-1}\mathbf{v} = \mathbf{v}', \qquad \rho^{-1}U^{-2}p = p'. \tag{4.2.5}$$

For the moment we assume that $\mathbf{F} = 0$. Then, by substituting the above relations in to the dynamical equation, we get:

$$\frac{\partial \mathbf{v}'}{\partial t'} + \mathbf{v}' \cdot \nabla' \mathbf{v}' = -\nabla' p' + \frac{\nu}{LU} \nabla'^2 \mathbf{v}'. \tag{4.2.6}$$

Thus, we naturally derive a useful dimensionless quantity, the Reynolds number:

$$\mathscr{R} = \frac{LU}{\nu}. \tag{4.2.7}$$

The dimensional quantities L and U are generally chosen so that they reflect the global geometry, such as the size of the apparatus, and the mean flow of the fluid. The Reynolds number measures the relative strength of the linear term and the non-linear interaction. The Reynolds number is a single parameter which determines the character of the flow in a particular geometry and its value can be used to measure the onset of turbulence. This suggest the following physical interpretation:

$$\mathscr{R} \approx \frac{\text{inertial forces}}{\text{viscous forces}}. \tag{4.2.8}$$

When viscous forces are dominant the flow is smooth. Such flows can be well treated analytically. As the Reynolds number increases the inertial forces dominate the dynamics. The smoothness of the flow is lost through the appearance of rapid and irregular fluctuations in the velocity field.

The equation of motion for electrically conducting fluids are more complicated because they must accommodate the effect of electromagnetic fields. The effect of electromagnetic interactions is fully captured by Maxwell's equations:

$$\nabla \times \mathbf{B} = \mu_0 \mathbf{j} + \frac{1}{c^2}\frac{\partial \mathbf{E}}{\partial t}, \tag{4.2.9}$$

$$\nabla \cdot \mathbf{B} = 0, \tag{4.2.10}$$

$$\nabla \times \mathbf{E} = -\frac{\partial \mathbf{B}}{\partial t}, \tag{4.2.11}$$

$$\nabla \cdot \mathbf{E} = \frac{\rho_p}{\epsilon_0}, \tag{4.2.12}$$

where $\mathbf{B}$ is the magnetic field, $\mathbf{E}$ is the electric field, μ_0 magnetic permeability, ϵ_0 permitivity of free space, c is the speed of light, and ρ_p is the charge density. Further, we would like to make some simplifying assumptions. We only consider electromagnetic variations which are non-relativistic [2], which affects the form of Ampere's law. From Eq. (4.2.11) we have $E/L \approx B/T$, where, as before, the quantities denote typical values. Using this we can asses the magnitude of the displacement current, which gives:

$$\frac{E}{c^2 T} \approx \frac{V^2}{c^2} |\nabla \times \mathbf{B}|, \tag{4.2.13}$$

where V is the ratio L/T. It follows that when electromagnetic variations are much smaller that the speed of light, the displacement current is negligible.

Plasmas which move with non-relativistic speed are subject to magnetic and electric fields, which exert force on the moving fluid. This relationship is captured by Ohm's law:

$$\mathbf{j} = \sigma (\mathbf{E} + \mathbf{v} \times \mathbf{B}), \tag{4.2.14}$$

where $\mathbf{j}$ is the current density and σ is the electric conductivity. The above two equations are now sufficient to derive the induction equation:

$$\frac{\partial \mathbf{B}}{\partial t} = \nabla \times (\mathbf{v} \times \mathbf{B}) - \frac{1}{\sigma \mu_0} \nabla \times (\nabla \times \mathbf{B}). \tag{4.2.15}$$

Application of vector identities and the solenoidal property of the magnetic field finally lead to the expression:

$$\frac{\partial \mathbf{B}}{\partial t} = \nabla \times (\mathbf{v} \times \mathbf{B}) + \eta \nabla^2 \mathbf{B}, \tag{4.2.16}$$

where $\eta = 1/\mu_0 \sigma$ is the magnetic diffusivity. If of interest, quantities such as $\mathbf{j}$ and $\mathbf{E}$ can be determined once the problem is solved for the magnetic and velocity fields.

We now have to consider the form of the force $\mathbf{F}$ in Eq. (4.2.1), which arises from the electromagnetic interaction of the fluid element. The force exerted on the charge q_i is the Lorentz force $q_i (\mathbf{E} + \mathbf{v}_i \times \mathbf{B})$. The force exerted on the fluid element is the sum of forces exerted on each charge in the fluid element. A fluid element is a macroscopic quantity and at this point we make another assumption that the plasma is quasi-neutral. This means that the fluid element is effectively neutral and we can ignore the effect of $\mathbf{E}$ because the total charge of the fluid element is approximately zero, $\delta q \approx 0$. It follows that the Lorentz force acting on the fluid element is:

$$\mathbf{F}_L = \mathbf{j} \times \mathbf{B}, \tag{4.2.17}$$

Bringing all the equations together we end up with a set of four equations:

$$\frac{\partial \mathbf{v}}{\partial t} = -\frac{1}{\rho}\nabla p + \nu \nabla^2 \mathbf{v} - \mathbf{v} \cdot \nabla \mathbf{v} + \frac{1}{\rho \mu_0} (\nabla \times \mathbf{B}) \times \mathbf{B}, \tag{4.2.18}$$

$$\frac{\partial \mathbf{B}}{\partial t} = \nabla \times (\mathbf{v} \times \mathbf{B}) + \eta \nabla^2 \mathbf{B}, \tag{4.2.19}$$

$$\nabla \cdot \mathbf{v} = 0, \tag{4.2.20}$$

$$\nabla \cdot \mathbf{B} = 0. \tag{4.2.21}$$

These are our defining equations of MHD. At this point it is appropriate to introduce new dimensionless quantities which are absent in hydrodynamics. When the induction equation is re-written in a dimensionless form, the relative strength of non-linear and linear terms is weighted by the magnetic Reynolds number:

$$\mathscr{R}_m = \frac{LV}{\eta}. \tag{4.2.22}$$

Clearly, a combination of dimensionless number is also a dimensionless quantity. One such combination which may be of interest is the magnetic Prandtl number, which is the ratio of the magnetic Reynolds number to the Reynolds number:

$$\mathscr{P} = \frac{\nu}{\eta}. \tag{4.2.23}$$

This quantity measures the relative effects of viscous versus magnetic diffusion rates. The importance of these dimensionless quantities in analytical studies is two fold: various limits can simplify the defining equations, for example note that, in the limit when $\mathscr{R}_m \ll 1$, the induction equation reduces to the diffusion equation and $\mathscr{R}_m$ can be used as an expansion parameter in perturbation treatments. Further, in computational studies, for example in hydrodynamics, the Reynolds number provides the scale of the mesh resolution for accurate modeling [6].

Our further analysis will be centered around Eq. (4.2.1) and Eqs. (4.2.18)–(4.2.21). Eventually we will introduce random forcing into Eqs. (4.2.18) and (4.2.19) to model sustainable turbulence, but the motivation for that will become clear from our later discussion on the renormalization group and its applications to hydrodynamics and MHD. To this end we have simply introduced the MHD equations. Various methods of analysis is the subject of the next section.

4.3 Methods to Study Turbulence

In this section we aim to briefly touch upon various methods used to date to analyze turbulence. Most of the discussion will be in the context of the NS equation. Because this equation has been studied to a much greater extent, an elementary presentation

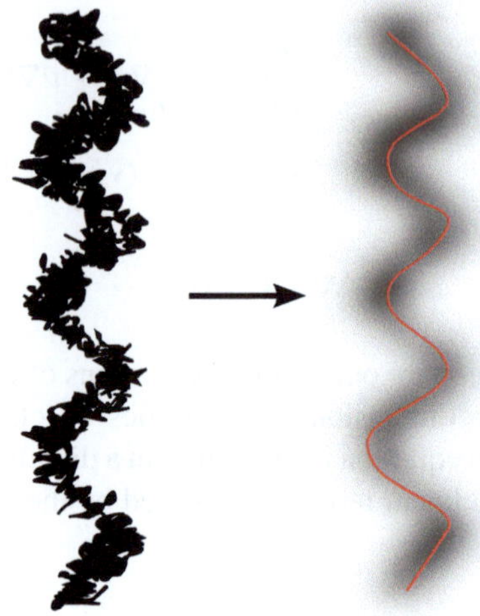

Fig. 4.1 Consider some complicated path depicted on the *right*. The *black line* represents the trajectory in space traced out by a hypothetical particle. The trajectory is highly irregular and the small-scale details are analogous to the eddies in a turbulent flow. If such a trajectory is reproducible, like in a stationary jet flow, these paths can be superimposed to give a smoother trajectory such as the one shown on the *right*. The *red line* represents a precise smooth trajectory, while the cloud represents statistical variations from the *red line*

of various methods can be more compactly presented using the NS equation only. Extensions to MHD are often rather straightforward but the algebraic form of the resultant equations becomes much more complicated. The various topics discussed in this section by no means form an exhaustive list in the field of turbulence studies. On the contrary, it can be viewed as rather subjective. As guiding principles, we have selected topics which, on an elementary level, can give either useful and important results, or present the problem in a new light.

4.3.1 Closure Modeling

Solving the NS equation with realistic boundary conditions in fully developed turbulence has proved to be a very difficult task. The profound problem lies in the non-linear interaction of many degrees of freedom across all length and time scales. One is then naturally led to look for other ways to quantify the complexity of the exact solutions. For instance, if we have a look at the flows from jets, it is apparent that there is some detail on every observable scale in the form of ever smaller eddies.

Perhaps, sometimes this is too much information for practical purposes and maybe we can get away with a more blurred image of these flows. Thus, we wish to sacrifice detail for solvability. Inevitably we are led to invoke a statistical description. The irregularities in the flow occur at random and, therefore, once reproducible, can be subject to a statistical description. In the case of stationary jet flows we can take multiple copies of the flow at different times and compute the average [6]. The average flow will be a smooth function with a better hope for an analytical description (Fig. 4.1). In practice, however, this is not very useful, since the main task of any theory is to be able to make predictions without a priori knowledge of the exact flow. In other words, we would like to start with the full NS equation and try to model the behavior of the averaged flows from it. A natural starting point is to break down the velocity field into the average/mean flow and fluctuations:

$$\mathbf{v}(\mathbf{x}, t) = \mathbf{U}(\mathbf{x}, t) + \mathbf{u}(\mathbf{x}, t), \tag{4.3.1}$$

where $\mathbf{U}$ is the average flow and $\mathbf{u}$ are fluctuations. Upon substitution into the NS equation we obtain:

$$\frac{\partial(\mathbf{U}+\mathbf{u})}{\partial t} + (\mathbf{U}+\mathbf{u}) \cdot \nabla(\mathbf{U}+\mathbf{u}) = -\frac{1}{\rho}\nabla(P+\tilde{p}) + \nu\nabla^2(\mathbf{U}+\mathbf{u}). \tag{4.3.2}$$

If we now take the average of the above equation we have:

$$\frac{\partial \mathbf{U}}{\partial t} + \mathbf{U}\cdot\nabla\mathbf{U} + \langle\mathbf{u}\cdot\nabla\mathbf{u}\rangle = -\frac{1}{\rho}\nabla P + \nu\nabla^2\mathbf{U}, \tag{4.3.3}$$

which follows from the fact that $\langle\mathbf{U}\rangle = \mathbf{U}$ and $\langle\mathbf{u}\rangle = 0$ by definition. The equation now has an extra term $\langle\mathbf{u}\cdot\nabla\mathbf{u}\rangle$, which requires a knowledge of a higher order moment of the unknown probability distribution. To this end we have one equation and two unknowns in the form of the average field and the second moment. Any attempt to derive the dynamics of the $\langle\mathbf{u}_i\mathbf{u}_j\rangle$ will inevitably introduce a third moment and the problem will persist. This is known formally as a closure problem [7]. Its origin lies in the non-linear term of the NS equation and is a general feature of any non-linear equation.

One possible solution to the problem is to postulate relations between statistical moments, since in principle an equation of the form

$$\mathbf{U} = f\left(\langle\mathbf{u}_i\mathbf{u}_j\rangle\right), \tag{4.3.4}$$

would render the set of equation closed. Such closure relations are neither exact nor derivable and can have a very limited range of applications. Such attempts to solve turbulence problem are usually met in engineering applications. With growing computer power more elaborate models emerge and it remains an active field of research [6].

4.3.2 Direct Numerical Simulations

An alternative to closure modeling is to try to solve the NS equation directly, such methods are known as direct numerical simulation. As can be anticipated this is a very challenging task, since turbulent flows span across large temporal and spacial scales. As with any direct simulation method it requires the system to be discretized on a grid or a mesh. It should be small enough to capture the smallest eddies in the flow, yet big enough to incorporate the largest eddies [6]. By calling these two scales η and L respectively, the number of grid point should roughly satisfy:

$$N \gg L/\eta. \tag{4.3.5}$$

This ratio of length scales is in fact related to the Reynolds number $L/\eta \approx \mathscr{R}^{3/4}$. By including appropriate resolution across time scales the overall expense of modeling *3D* turbulence goes as $\mathscr{R}^{11/4}$ which, considering that turbulent flows are characterized by large Reynolds number of order 10^6 and larger, such modeling is computationally expensive. Despite the expense of such computations, current numerical simulations can handle large numbers of grid points for isotropic and homogeneous turbulence [6]. Another feature of direct numerical simulation, not necessarily in the context of turbulence, is numerical dissipation or heating. This is an effect due to numerical approximations which artificially dissipates energy from the system. Sometimes such effects are favorable in a simulation, since the associated discretization methods are more stable. However, in the context of turbulence this effect should be minimized, since high numerical dissipation can over-damp flow fluctuations, leading to the spurious decay of turbulence [6].

Over the past decades with the development of computer power numerical simulations have become an indispensable tool in turbulence research and often is the only way to check the validity of analytical models. However, numerically attainable Reynolds numbers are still too small to understand real turbulence and, thus, studies of high Reynolds number turbulence are of a more academic nature.

4.3.3 Analytical Studies

4.3.3.1 Dynamical systems

Perhaps a rather natural framework to study turbulence is in the context of dynamical systems, since the NS equation falls precisely into the class of problems addressed by this field [8]. Conceptually this is very helpful, as the emergence of unpredictability of a deterministic equation can be explained through chaos. For an illustrative example consider a toy model for the NS equation:

$$v_{t+1} - v_t = -2v_t^2 - v_t + 1, \quad v_0 = \omega, \quad t = 0, 1, 2, \ldots. \tag{4.3.6}$$

The first, second and third terms of the right-hand side can be associated with a non-linear term, the viscous term and the forcing of the true NS equation respectively. The important property of this map is that it displays sensitivity to the initial condition which is a reminiscent of chaotic behavior. Furthermore, this equation is simply a fancy way to write the so called tent map, which is defined as follows:

$$x_{t+1} = \begin{cases} 2x_t & \text{for } 0 \leq x \leq \frac{1}{2} \\ 2 - 2x_t & \text{for } \frac{1}{2} \leq x \leq 1. \end{cases} \tag{4.3.7}$$

This has another important property, namely, the invariance of the measure (Lebesgue). For if, we consider the initial condition x_0 to be uniformly distributed, then the image of the tent map will also follow a uniform distribution. The fact that the initial condition is chosen at random rather than being deterministic is almost irrelevant as a result of Birkhoff's ergodic theorem, which states that the time average of some integrable function along the trajectory from a given initial condition is equal to the ensemble average with the invariant measure.

The lesson to take from the above oversimplified example is this: the analysis of dynamical systems provides a natural framework for a statistical description of turbulence and exploits chaos to explain the non-predictability of the flow. However, apart from these qualitative features little has been done in the field to understand turbulence. The application of traditional machinery in the form of Lyapunov exponents, attractors etc. is hard to implement for systems with many degrees of freedom such as the NS equation [9, 10]. This has served as the main obstacle to progress in this direction [8].

4.3.3.2 Phenomenological Models

Phenomenological theories are based on non-rigorous plausible hypotheses, together with dimensional arguments, to analyze turbulence. Perhaps surprisingly, such an approach has given one of the greatest insights into the nature of turbulence. This methodology allows one to calculate energy spectrum functions, count degrees of freedom etc. Considerable work on phenomenology is based on the work of Kolmogorov [11, 12]. These studies were primarily concerned with hydrodynamics, with the arguments being extended later to MHD. In this section we will outline only a few results, while more complete reviews can be found in the original papers or standard textbooks [7, 8, 13, 14].

Together with statistical insight, dimensional analysis can be used to derive the energy spectrum in turbulent flows. The idea of energy cascades was first put forward by Richardson [6]. The mechanism of energy transfer in turbulent flows is explained via local interactions of large eddies, which carry the kinetic energy down to the smallest eddies, where energy is dissipated through viscous forces. The intermediate range between the largest eddies and the smallest eddies is called the inertial range. This idea can be exploited further, for if we make two assumptions, namely, scale-invariance within the inertial range, and strictly local interactions, then we could argue that, in fully developed turbulence, we can identify two relevant length scales: a typical size of the largest eddies, δ, which is comparable to the system size, while viscosity sets a lower bound η on the size of the smallest eddies. Let us assume that $\eta \ll \delta$. Large-scale eddies are generated by the shear stresses in the flow and the small scales are being dissipated by viscosity. It is reasonable to conclude that there is a mechanism through which there is an energy transfer from large scales, where energy is being produced, to small scales, where it is dissipated. Moreover this would require an intermediate or inertial range where energy is neither produced or dissipated. Then, in equilibrium, the energy flux through the inertial range should be equal to the rate at which the energy is dissipated. Hence, the energy cascades across the inertial range, from large to small scales.

The inertial range is shared by both the large-scale energetic regime and the small-scale dissipative regime. Large-scale eddies are not directly affected by viscosity. Hence, since the inertial and energetic regions overlap, the energy distribution cannot be affected by viscosity. On the other hand, small-scale eddies are universal, all the information about the original geometry is lost, so in the proximity of η, the inertial range cannot depend on the large-scale structure δ. The energy distribution across different size eddies in the inertial range should in principle depend on all relevant scales of the flow:

$$E = E\left(\delta, \nu, r, \varepsilon\right), \tag{4.3.8}$$

where ε is the rate of the energy dissipation and r is the typical eddie size. However, as we have argued, the large-scale and small-scale structures should not influence the inertial range, thus:

$$E = E\left(r, \varepsilon\right). \tag{4.3.9}$$

On the grounds of dimensional analysis we can postulate the simplest possible functional relation between the two such that the result has dimension of energy. It follows that the energy spectrum is:

$$E \propto (\varepsilon r)^{2/3}. \tag{4.3.10}$$

This is known as Kolmogorov's energy spectrum. This is more commonly written as a $-5/3$ law in Fourier space. Today it is believed that Kolmogorov's spectrum is correct [6].

In the 1960s, the first phenomenological model of MHD turbulence was put forward by Kraichnan [18]. He proposed a different exponent to Kolmogorov's spectrum, namely:

$$E\left(k\right) \propto k^{-3/2}. \tag{4.3.11}$$

This was attributed to an interaction of localized modes with an external magnetic field, which is irrelevant in hydrodynamics.

Kolmogorov's energy spectrum is a very powerful result. However, it is immediately evident that phenomenological models cannot account for the full description of turbulence. To determine proportionality constants one has to fall back onto experiments and the theory offers no means of computing them explicitly. As such, phenomenological models must be used in combination with other analytical methods.

4.3.3.3 Functional methods

Turbulent flows are necessarily statistical in nature. Thus, it is natural to consider probability distribution functions which govern such flows. This leads us to the concept of functional integration methods, which are concerned with averaging over such probability measures in function space. This functional formalism was initiated by Hopf [15] when he derived the governing equation of the probability functional, known as the Hopf equation for the NS equation. The Hopf equation enabled the evaluation of correlation functions at a single moment in time [16]. Later, the method was generalized to compute multi-point correlation function at different times. This is based on work of Lewis and Kraichnan [17] and, in view of its generality, we outline their methodology.

Consider the generating functional $\Phi\left[\mathbf{Z}\left(\mathbf{x},t\right)\right]$ of a probability distribution $P\left[\mathbf{v}\left(\mathbf{x},t\right)\right]$:

$$\Phi\left[\mathbf{Z}\left(\mathbf{x},t\right)\right] = \left\langle \exp\left(i\left[\mathbf{Z}\cdot\mathbf{v}\right]\right)\right\rangle, \tag{4.3.12}$$

where the dot product denotes:

$$\mathbf{Z}\cdot\mathbf{v} \equiv \int d\mathbf{x} \int dt \mathbf{Z}\left(\mathbf{x},t\right)\cdot\mathbf{v}\left(\mathbf{x},t\right). \tag{4.3.13}$$

Functional differentiation of $\Phi\left[\mathbf{Z}\left(\mathbf{x},t\right)\right]$ with respect to $Z_\alpha\left(\mathbf{x}',t'\right)$ generates an average of the velocity field over the probability distribution functional:

$$\left.\frac{\delta\Phi\left[\mathbf{Z}\left(\mathbf{x},t\right)\right]}{\delta Z_\alpha\left(\mathbf{x}',t'\right)}\right|_{\mathbf{Z}=0} = i\left\langle v_\alpha\left(\mathbf{x}',t'\right)\right\rangle. \tag{4.3.14}$$

The goal is to write the evolution equation for the generating functional using the NS equation. Thus, consider the quantity:

$$\frac{\partial}{\partial t}\left\{\frac{\delta\Phi\left[\mathbf{Z}\right]}{\delta Z_\alpha\left(\mathbf{x},t\right)}\right\} = i\left\langle\frac{\partial v_\alpha\left(\mathbf{x},t\right)}{\partial t}\exp\left(i\left[\mathbf{Z}\cdot\mathbf{v}\right]\right)\right\rangle. \tag{4.3.15}$$

Using the NS equation, the above relationship can be written as:

$$\frac{\partial}{\partial t}\left\{\frac{\delta\Phi\left[\mathbf{Z}\right]}{\delta Z_\alpha\left(\mathbf{x},t\right)}\right\} = i\left\langle\left(-\frac{\partial\left\{v_\beta\left(\mathbf{x},t\right)v_\alpha\left(\mathbf{x},t\right)\right\}}{\partial x_\beta} - \frac{\partial p}{\partial x_\alpha} + \nu\frac{\partial^2 v_\alpha\left(\mathbf{x},t\right)}{\partial x_\beta\partial x_\beta}\right)\exp\left(i\left[\mathbf{Z}\cdot\mathbf{v}\right]\right)\right\rangle. \tag{4.3.16}$$

It follows that the above manipulations transform a non-linear problem in $\mathbf{v}$ to a linear problem in $\Phi\left[\mathbf{Z}\left(\mathbf{x},t\right)\right]$. The only task now is to express the pressure term as a functional derivative of $\Phi\left[\mathbf{Z}\left(\mathbf{x},t\right)\right]$. This can be readily achieved for the NS equation. The equation can be written as:

$$\frac{\partial}{\partial t}\left\{\frac{\delta\Phi[\mathbf{Z}]}{\delta Z_{\alpha}(\mathbf{x},t)}\right\}=i\frac{\partial}{\partial x_{\beta}}\left\{\frac{\delta^{2}\Phi[\mathbf{Z}]}{\delta Z_{\alpha}(\mathbf{x},t)\,\delta Z_{\beta}(\mathbf{x},t)}\right\}+\nu\frac{\partial^{2}}{\partial x_{\beta}\partial x_{\beta}}\left\{\frac{\delta\Phi[\mathbf{Z}]}{\delta Z_{\alpha}(\mathbf{x},t)}\right\}-\frac{\partial\Pi}{\partial x_{\alpha}}. \tag{4.3.17}$$

Here, we follow [17]. Let us introduce a solenoidal dummy field $\mathbf{Y}$, which satisfies:

$$\frac{\partial Y_{\beta}(\mathbf{x},t)}{\partial x_{\beta}}=0, \tag{4.3.18}$$

and $Y_{\beta}(\mathbf{x},t)\rightarrow 0$ as $|x|\rightarrow\infty$. We multiply Eq. (4.3.17) by $Y_{\alpha}(\mathbf{x},t)$ and integrate over all space:

$$\begin{aligned}0=&\int d\mathbf{x}Y_{\alpha}(\mathbf{x},t)\left\{\frac{\partial}{\partial t}\left\{\frac{\delta\Phi[\mathbf{Z}]}{\delta Z_{\alpha}(\mathbf{x},t)}\right\}-i\frac{\partial}{\partial x_{\beta}}\left\{\frac{\delta^{2}\Phi[\mathbf{Z}]}{\delta Z_{\alpha}(\mathbf{x},t)\,\delta Z_{\beta}(\mathbf{x},t)}\right\}\right\}\\&-\int d\mathbf{x}Y_{\alpha}(\mathbf{x},t)\left\{\nu\frac{\partial^{2}}{\partial x_{\beta}\partial x_{\beta}}\left\{\frac{\delta\Phi[\mathbf{Z}]}{\delta Z_{\alpha}(\mathbf{x},t)}\right\}-\frac{\partial\Pi}{\partial x_{\alpha}}\right\}.\end{aligned} \tag{4.3.19}$$

Using integration by parts the last term vanishes:

$$\int d\mathbf{x}Y_{\alpha}(\mathbf{x},t)\frac{\partial\Pi(\mathbf{x},t)}{\partial x_{\alpha}}=-\int d\mathbf{x}\Pi(\mathbf{x},t)\frac{\partial Y_{\alpha}(\mathbf{x},t)}{\partial x_{\alpha}}, \tag{4.3.20}$$

because the dummy field $\mathbf{Y}$ is solenodial. Since the test field $Y_{\alpha}(\mathbf{x},t)$ is otherwise arbitrary, the equation should be valid for any realization of this field. It follows that the term in the brackets must satisfy:

$$\frac{\partial}{\partial t}\left\{\frac{\delta\Phi[\mathbf{Z}]}{\delta Z_{\alpha}(\mathbf{x},t)}\right\}-i\frac{\partial}{\partial x_{\beta}}\left\{\frac{\delta^{2}\Phi[\mathbf{Z}]}{\delta Z_{\alpha}(\mathbf{x},t)\,\delta Z_{\beta}(\mathbf{x},t)}\right\}-\nu\frac{\partial^{2}}{\partial x_{\beta}\partial x_{\beta}}\left\{\frac{\delta\Phi[\mathbf{Z}]}{\delta Z_{\alpha}(\mathbf{x},t)}\right\}=0. \tag{4.3.21}$$

By further differentiation with respect to Z_{γ} a hierarchy of cumulants can be derived.

The sole purpose of the above reformulation is to present the problem of turbulence in a new light. To this end the Hopf equation serves as an elegant and rigorous formulation of the problem. It can also provide new approximation schemes which are not evident from Eq. (4.2.1) [8].

4.3.3.4 Renormalized Perturbation Theories and Direct Interaction Approximation

Perturbation methods have been widely used in the context of statistical mechanics when one is faced with an equilibrium system with weak interactions. A primitive perturbation series then consists of an expansion with a weak coupling as an expansion parameter. However, one is often interested in the behavior of a system in the thermodynamic limit where the number of particles N and the volume of the system V both go to infinity, while the ratio N/V is kept constant. As an example, for a dilute neutral gas [19] which interacts via a Lennard-Jones potential, the thermodynamic limit results in an inhomogeneous expansion, namely the expansion parameter to a given order has a variety of different dependences on N/V. To overcome this problem, one effectively re-shuffles the perturbation theory to express it as a power series in N/V. Then, by assuming a low density regime, such a series can be truncated at low orders and systematic corrections can be made. In practice this amounts to summing over an infinite sub-series of the original expansion. This formalism naturally leads to the introduction of graphs as means of representing algebraic terms in a series whose topological structure is in one-to-one correspondence with the algebraic structure of a particular term.

The extension of the above methods to turbulence is a non-trivial task. Turbulence is a non-equilibrium many-body problem with strong interactions. It is characterized by a high Reynolds number, so any primitive expansion in this parameter would necessarily lead to strongly divergent series. As an attempt to improve the behavior of such a perturbation expansion, one considers so-called renormalized perturbation theory. In effect this amounts to a re-arrangement of the initial series. We do not wish to go into technical details here, in part since they are quite involved and eventually will closely resemble our own calculation. Some of the original work on the subject can be found in Refs. [20, 21].

The technique of renormalized perturbation theories amount to the following steps. Starting with a randomly stirred NS equation, an expansion in the non-linear term is performed. This generates a series expansion for $\mathbf{v}$. Quantities of interest are correlation and response functions. Using the initial expansion for $\mathbf{v}$ one can construct a corresponding series expansion for these functions as well. By re-expressing the series in terms of graphs, a structure emerges, namely, that the exact series expansion for either the correlation or the response functions can be formed by re-defining/renormalizing the bare propagator and the vertex, which are the building blocks for the former series. This fact substantially reduces the number of terms/diagrams required in the description. However, this is simply a method and does not constitute a physical theory. The advantage of the above procedure is that the initial divergent series has been replaced by one with unknown properties. It might be convergent or not, or be asymptotic. Inevitably, the series for the renormalized quantities still have to be truncated at a point where it is still analytically tractable.

A particular truncation scheme which has been used in the context of hydrodynamic turbulence, as well as MHD, is known as the direct interaction approximation (DIA). This pioneering work was due to Kraichnan [18]. The approximation amounts to going to the second order in the number of verticies, with the propagator being renormalized to second order as well. The three-point vertex is not renormalized in this scheme which, in effect, is the direct interaction approximation. In the context of a perturbation expansion, this is an *ad hoc* truncation scheme.

An important consequence of the DIA is that it predicts an energy spectrum which follows a $-3/2$ power law. This is attributed to the non-local effects of sweeping, which are excluded from Kolmogorov's considerations. Today, it has been established experimentally that the inertial energy spectrum follows more closely the Kolmogorov $-5/3$ power law. Despite that, the DIA is still of considerable interest.

The major strength of the DIA lies in the fact that it is a physically realizable theory. In contrast, other approximation schemes lead to the unphysical prediction of a negative energy spectrum for the theory.

Renormalized perturbation theories are in a good agreement with simulations at low Reynolds number [8]. However, the application of these methods to inhomogeneous flows has posed great difficulty.

4.4 Application of Renormalization Group to Hydrodynamic and Magnetohydrodynamic Turbulence

Following success of RG method in the theory of critical phenomena, applications extended to the study of evolution equations with appropriate extensions of RG group methods, which is known as the dynamic RG. Traditional RG applications were intended to compute scaling laws. However, as follows from Kolmogorov's phenomenological theory, in turbulence, scaling exponents can be computed based on simple dimensional and statistical arguments (in the inertial range). Thus, to this end the RG method, as applied to turbulence, should offer something more in order to justify itself. For example, Kolmogorov's energy spectrum is primarily a proportionality relation, so if the RG method is to make some non-trivial predictions it must predict such coefficients.

The mathematical complexity of treating the full NS equation with realistic boundary conditions has proved to be extremely hard. A lot of present research is centered around a somewhat simpler problem, namely, statistically homogeneous and isotropic turbulence. All of the discussion which follows will be in the context of this simpler problem. We will focus on unbounded fluids. Thus, to sustain turbulent flows, random stirring forces are introduced, which play the role of an energy input and provide a statistically stationary state. The statistics of the stirring forces is often chosen to follow a Gaussian distribution, such that the second moment is white-in-time and follows a power-law spectrum. The power-law form of the spectrum is needed to establish the self-similarity property of the stirring forces which are used in RG methods, with further specification subject to individual studies. With these specifications the NS equation takes a form of a Langevin equation.

Langevin equations have proved to be very useful in many physics applications. The most elementary use of these equations goes back to Einstein [22] in 1905 and his quantitative theory of Brownian motion. Application of RG methods to Langevin equations was initiated by Ma and Mazenko [23] in 1975. They employed essentially a Wilson-Kadanoff-style RG.

4.4.1 Hydrodynamics

In 1977, following Ma and Mazenko [23], Forster and others [24] applied RG methods to the randomly stirred NS equation. All of the above considerations used to simplify the problem apply to their work, and all other problems discussed here, unless stated otherwise. They considered a number of different spectra which specify the stirring forces. To model the fluid near equilibrium, often referred to as Model A, the forcing spectrum has the form:

$$D_A(k) = \begin{cases} D_0 k^2 & \text{for } |k| < \Lambda, \\ 0 & \text{otherwise,} \end{cases} \tag{4.4.1}$$

where Λ is the short scale cut-off. This forcing can be associated with molecular randomness in the flow. In addition they considered another spectrum, known as Model B:

$$D_B(k) = \begin{cases} D_0 & \text{for } |k| < \Lambda, \\ 0 & \text{otherwise.} \end{cases} \tag{4.4.2}$$

Note, that when $k \to 0$ the spectrum is non-vanishing, $D_B(k) \nrightarrow 0$. The stirring persists to the size comparable to the system size. A physical interpretation of this forcing spectra is to think of shaking the fluid as a whole. In a more realistic bounded flow, say fluid in a vessel, this would correspond to shaking the vessel. A third model, referred to as Model C, corresponds to a spectrum:

$$D_C(k) = \begin{cases} D_0 & \text{for } \bar{\Lambda} < |k| < \Lambda, \\ 0 & \text{otherwise.} \end{cases} \tag{4.4.3}$$

This forcing implies that energy is injected only in a certain spatial range, which corresponds to $(\bar{\Lambda}, \Lambda)$ in Fourier space. The objective of this paper is to study the long-time properties of the correlation function subject to different forcing.

For model A, it has been found that a non-trivial fixed point exists in $d > 2$. Analysis near the fixed point shows that long-time properties of the viscosity are in agreement with previous literature. In $d = 2$, there is a logarithmic correction to conventional hydrodynamics. The energy spectrum function is found to be $E(k) \propto k^{d-1}$. In $d < 2$, the hydrodynamic description breaks down, though it is unclear what this model could be below $d = 2$ in the first place. A similar analysis is carried out for model B. It is found that a hydrodynamic description breaks down for $d < 4$. Logarithmic corrections follow at $d = 4$, and the long-time properties of viscosity can be computed for $d > 4$. The energy spectrum is not the same as in Model A and is very different depending on dimensionality of the problem. This is associated with the shaking aspect of the fluid. An important result of this paper is that Model A and C in fact display the same asymptotic behavior. Note, model C can be thought of as a type of Model B, where the consecutive elimination of degrees of freedom is partial. In this case the forcing spectrum is dominated by k^2 behavior in the limit of small k and, therefore, closely resembles model A and is attributed to universality in the sense we have discussed in the previous chapter.

In 1979, de Dominicis and Martin, published their paper [25] on asymptotic properties of the long-wavelength fluctuations using field-renormalization-group techniques. This calculation closely resembles methods applied in QFT. They have used the following form of the energy spectrum:

$$D(k) \propto D_0 \frac{k^{4-d}}{\left(m_0^2 + k^2\right)^{-y/2}}. \tag{4.4.4}$$

The quantity m_0 plays a role of the infrared cut-off of the stirring forces. The forcing domain is $m_0 \ll k \ll \Lambda$. The cut-off, Λ, is introduced at the intermediate stages of the calculation but eventually is send to infinity. The results of [24] are recovered in this framework. In addition, Kolmogorov's energy spectrum is recovered by an appropriate choice of y, namely $y \to 4^-$, and $d > 2$. Although this has been very reassuring, the central question of why this particular form of energy spectrum should be observed in experiments has not been addressed, as was stressed by the authors themselves. So far there has been no attempt to predict the proportionality coefficients for the scaling laws. In fact, by this time it was argued by Kraichnan [21], that RG methods are not superior to any other closure schemes. As we have pointed out, the ability to predict the energy spectrum rests solely on dimensional grounds. Perhaps, one could argue that at this stage RG provided at least a conceptual framework to understand universality, as was demonstrated in [24].

To counter Kraichnan's point, in 1983 Fournier and Frisch [26] carried out a calculation, this time Wilson-Kadanoff style, to relate Kolmogorov's constant to the energy input rate of the random force. The coefficient is shown to have a universal property, the context being that they are independent of viscosity and small-scale forcing. Following this, in 1986 Yakhot and Orszag [27] exploited RG applications even further. There are two sides to their work: on one hand, they have used RG methods to compute proportionality constants (explicit numerical values), which are in a reasonable agreement with some experiments; on the other hand, as it appears from numerous articles [28–31] which review their work, that there are many flaws with regards to their use of the RG. An important ingredient of their analysis is the ϵ-expansion. This is a small parameter which is introduced through the noise spectrum exponent. By choosing $\epsilon \to 0$, the effect of additional interactions which are produced by the RG can be studied. They have shown that it is possible to neglect some non-linear effects which are generated by the RG in this limit. However, to consistently reproduce Kolmogorov's spectrum it is required that $\epsilon = 4$. Thus, a simple self-consisted analysis seems to be feasible around $\epsilon \to 0$, but the physics dictates that the appropriate regime is around $\epsilon = 4$. Extrapolation from $\epsilon \to 0$ to $\epsilon = 4$ is not obvious. Further, Eyink [28] argued that even in the limit $\epsilon \to 0$, those terms which were termed to be irrelevant, are in fact marginal by power counting.

As review articles suggest [20], all of the above issues are still open. Our purpose is to investigate these matters for ourselves in the slightly more generalized framework of MHD turbulence. To this end, RG applications to MHD are even more controversial.

4.4.2 Magnetohydrodynamics

We begin our discussion of the renormalization group analysis of magnetohydrodynamics with a 1981 paper by Fournier et al. [32]. The authors used a dynamic renormalization group technique [23] and worked with the original field variables, $\{\mathbf{v}, \mathbf{B}\}$. Note that this is a new feature of MHD equations, since one can choose to work with Elsasser variables [33] instead. The stochastic driving of the kinetic and magnetic equations are assumed to be independent, zero-mean and Gaussian random functions, with different exponents. Accordingly, they identified regimes under which the RG procedure generates k^2 corrections to the noise covariance. The relevance/irrelevance of these corrections is discussed. Another important result of their work is the fact that the vertex associated with the Lorentz force in the kinetic equation is renormalized, while other vertices are not. In dimensions $d > 2$ they identified two non-trivial regimes. Vertex renormalization plays an important part in their analysis and leads to predictions which differ from other analytical techniques used in the study of MHD turbulence (the direct interaction approximation, for example). Further, the vertex renormalization is an additional complexity which is not present in NS turbulence. The NS equations are invariant under Galilean transformations, which prevents renormalization of the vertex. However, as has been shown by McComb and Berrera [34, 35] a more careful analysis of these ideas is required. In [35] vertex invariance under RG transformations was shown using Ward-Takahashi identities. This demonstrates an advantage of working in a functional integral representation of a stochastic partial differential equation.

A paper by Camargo and Tasso [36] used a slightly different approach by working with Elsasser variables. By the virtue of their approach, they weighted the Lorentz force and the inertial term in the same way. Consequently, they derive the renormalization flow equations for the couplings, kinematic viscosity and resistivity. An investigation of the flow equations in two and three dimensions is given with a focus on asymptotic properties of the couplings and the Prandtl number. Our main criticism of this work stems from the fact that no discussion is given about the vertex and noise renormalization. In this sense their work is reminiscent of the DIA approach of truncating the series expansion. Further, we would expect that regardless of how the non-linearities are weighted, the differential RG equations should reduce to the NS differential RG equation, which we could not establish. Also, as has been established for the NS case [28], the RG procedure generates new terms and the only systematic way of establishing their effect is by using scaling relations near a fixed point. The problem of marginal variables is clearly established for the NS case and it seems natural that an adequate discussion of such terms, for the MHD case, should be present. After all the NS equation is a subclass of the MHD equations.

More recent papers on the subject include [37]. The formal set-up of the problem is identical to [36], however the RG treatment is performed using the functional integral formalism. The RG calculation is carried out by successive integration over the small scales i.e a Wilsonian style calculation. Arguably, this is a more natural framework for the RG methods, since other field theoretic tools such as Ward-Takahashi identities,

Slavnov–Taylor identities, etc. can be conveniently derived [38]. In this work the RG equations derived are different to those in [36]. This already raises some questions about the validity of the analysis performed in the three papers discussed above. The resultant equations do reduce to the known results in NS studies. Further, the exponent, which is used to define noise correlation function, is adjusted to fit either the Kolmogorov or the Kraichnan energy spectrum in the vicinity of a non-trivial fixed point. However, the purpose of doing so is not clear. The RG procedure is a local transformation of equations and cannot account for any non-local effects such as sweeping. A discussion about vertex renormalization is given, where it is found that it is not renormalized. To one-loop order, a regime is chosen where the noise covariance is not renormalized and there is no discussion about the characterization of new terms which are generated at one-loop order.

In [39] a classical field-theoretic treatment of MHD turbulence has been given through the identification of the primitively divergent vertex functions. The study was restricted to two dimensions and the existence of a non-trivial fixed point was established. Further shortcomings in RG applications to MHD turbulence were reported by [40]. They have managed to reproduce the RG coefficient functions to one-loop order, which can be conveniently reduced to those equations obtained via the Wilsonian RG [32]. A kinetic fixed point, which is associated with the Kolmogorov scaling regime, was identified for $d \geq 2$.

We believe that, in view of the foregoing remarks, a consistent treatment of MHD turbulence is lacking in the framework of Wilsonian RG. This serves as a starting point for our own work.

4.5 Functional Integral

4.5.1 Elsasser Transformation

4.5.1.1 Dynamic Equations

The equations which describe MHD have been derived in the previous section. By using standard vector identities the defining equation can be written in a different form that is more convenient for subsequent analysis. They form a set of coupled stochastic partial differential equations:

$$\frac{\partial \mathbf{v}}{\partial t} + (\mathbf{v} \cdot \nabla)\, \mathbf{v} = -\frac{1}{\rho} \nabla P + \frac{1}{\eta \rho} \left((\mathbf{B} \cdot \nabla)\, \mathbf{B} - \frac{1}{2} \nabla\, |\mathbf{B}|^2 \right) + \nu \nabla^2 \mathbf{v} + \xi, \tag{4.5.1}$$

$$\frac{\partial \mathbf{B}}{\partial t} = (\mathbf{B} \cdot \nabla)\, \mathbf{v} - (\mathbf{v} \cdot \nabla)\, \mathbf{B} + \mu \nabla^2 \mathbf{B} + \Psi. \tag{4.5.2}$$

It is convenient to re-write them in a more symmetric form using Elsasser variables [33]. This is a two-step transformation. First, we rescale the magnetic, pressure and the magnetic noise fields:

$$\mathbf{b} = \frac{\mathbf{B}}{\sqrt{\rho\eta}}, \quad p = \frac{P}{\sqrt{\rho}}, \quad \psi = \frac{\Psi}{\sqrt{\rho\eta}}. \tag{4.5.3}$$

This transformation sets the magnetic and the velocity fields on the same footing in terms of their dimensionality:

$$\frac{\partial \mathbf{v}}{\partial t} + (\mathbf{v} \cdot \nabla)\,\mathbf{v} = -\nabla p + (\mathbf{b} \cdot \nabla)\,\mathbf{b} - \frac{1}{2}\nabla\,|\mathbf{b}|^2 + \nu\nabla^2\mathbf{v} + \xi, \tag{4.5.4}$$

$$\frac{\partial \mathbf{b}}{\partial t} = (\mathbf{b} \cdot \nabla)\,\mathbf{v} - (\mathbf{v} \cdot \nabla)\,\mathbf{b} + \mu\nabla^2\mathbf{b} + \psi. \tag{4.5.5}$$

The second step is to introduce the transformation:

$$\mathbf{P} = \mathbf{v} + \mathbf{b}, \quad \mathbf{Q} = \mathbf{v} - \mathbf{b}, \tag{4.5.6}$$

which leads to the equations:

$$\frac{\partial \mathbf{P}}{\partial t} + (\mathbf{Q} \cdot \nabla)\,\mathbf{P} = -\nabla\left(p + \frac{1}{2}\,|\mathbf{b}|^2\right) + \gamma_+\nabla^2\mathbf{P} + \gamma_-\nabla^2\mathbf{Q} + \xi + \psi, \tag{4.5.7}$$

$$\frac{\partial \mathbf{Q}}{\partial t} + (\mathbf{P} \cdot \nabla)\,\mathbf{Q} = -\nabla\left(p + \frac{1}{2}\,|\mathbf{b}|^2\right) + \gamma_-\nabla^2\mathbf{P} + \gamma_+\nabla^2\mathbf{Q} + \xi - \psi, \tag{4.5.8}$$

where we have defined:

$$\gamma_\pm = \frac{1}{2}\,(\nu \pm \mu)\,. \tag{4.5.9}$$

We want to eliminate the pressure term so the equation is expressed solely in terms of **P** and **Q** fields. It is useful to define a scalar function:

$$g = p + \frac{1}{2}\,|\mathbf{b}|^2\,. \tag{4.5.10}$$

If we take divergence of the MHD equations in Elsasser variables, all the linear terms in the field vanish, because of the incompressibility condition and Maxwell's equation. This leaves us with a relationship:

$$\nabla \cdot ((\mathbf{Q} \cdot \nabla)\,\mathbf{P}) = -\nabla^2 g, \tag{4.5.11}$$

$$\nabla \cdot ((\mathbf{P} \cdot \nabla)\,\mathbf{Q}) = -\nabla^2 g. \tag{4.5.12}$$

In the index notation, the left-hand side of the equations are:

$$\partial_i \left(Q_k \partial_k P_i\right) = \partial_i \partial_k \left(Q_k P_i\right), \tag{4.5.13}$$

and

$$\partial_i \left(P_k \partial_k Q_i\right) = \partial_i \partial_k \left(Q_i P_k\right). \tag{4.5.14}$$

Hence, we can express the scalar function g formally as:

$$g = -\nabla^{-2} \left(\partial_i P_k\right) \left(\partial_k Q_i\right), \tag{4.5.15}$$

where ∇^{-2} is an inverse of the Laplacian operator. Thus the MHD equations can be written as:

$$\frac{\partial P_i}{\partial t} + \left\{\partial_j \delta_{ki} - \nabla^{-2} \partial_i \partial_j \partial_k\right\} P_k Q_j = \gamma_+ \nabla^2 P_i + \gamma_- \nabla^2 Q_i + \xi_i + \psi_i, \tag{4.5.16}$$

$$\frac{\partial Q_i}{\partial t} + \left\{\partial_k \delta_{ij} - \nabla^{-2} \partial_i \partial_j \partial_k\right\} P_k Q_j = \gamma_- \nabla^2 P_i + \gamma_+ \nabla^2 Q_i + \xi_i - \psi_i. \tag{4.5.17}$$

These equations are remarkably similar to the equations of hydrodynamics. The apparent symmetry of the fields allows for a simpler application of the diagrammatic technique.

4.5.1.2 Noise Statistics

The statistics of the noise fields is specified using moments of the functional probability distribution. Since we are dealing solely with Gaussian fluctuations it is sufficient to specify the first and the second moments of the distribution. Thus, we have:

$$\left\langle \xi_i \left(\mathbf{k}, \omega\right) \xi_j \left(\mathbf{q}, \Omega\right)\right\rangle = D\left(k\right) \delta_{ij} \delta\left(\mathbf{k} + \mathbf{q}\right) \delta\left(\omega + \Omega\right). \tag{4.5.18}$$

In general, the noise covariance D, can depend on powers of the gradient operator. We assume that the mean of the fluctuations is zero. However, the space of all possible functions which represent the noise has to be further reduced due to the condition of incompressibility:

$$\nabla \cdot \xi = 0. \tag{4.5.19}$$

This can be conveniently achieved through functional integral representation of the probability functional:

$$\mathscr{P}\left[\xi\right] \rightarrow \mathscr{P}_R\left[\xi\right] = \mathscr{P}\left[\xi\right] \delta\left[\nabla \cdot \xi\right]. \tag{4.5.20}$$

In other words we want to calculate:

$$\langle \xi_i (\mathbf{x}, t)\, \xi_j (\mathbf{y}, \tau) \rangle = \int \mathscr{D}[\xi]\, \xi_i (\mathbf{x}, t)\, \xi_j (\mathbf{y}, \tau)\, \mathscr{P}[\xi]\, \delta[\nabla \cdot \xi], \qquad (4.5.21)$$

The functional probability distribution is:

$$\mathscr{P}[\xi] \propto \exp\left(-\frac{1}{2} \int dx\, dy\, \xi(x)\, \Gamma^{-1}(x, y)\, \xi(y) \right), \quad x \equiv (\mathbf{x}, t), \qquad (4.5.22)$$

where Γ is the noise covariance matrix. In the formalism of functional integrals we do not keep track of field-independent normalization constants since we are primarily interested in average quantities such as:

$$\langle \cdot \rangle = \frac{\int \mathscr{D}[\xi]\, (\cdot)\, \mathscr{P}[\xi]}{\int \mathscr{D}[\xi]\, \mathscr{P}[\xi]}, \qquad (4.5.23)$$

where such normalization constants cancel. Even when they are infinite, as they often are in condensed-matter-theory problems, we can still ignore them. A more detailed analysis will be given in a later section.

For now we seek to compute the second moment for the reduced probability distribution defined above. At this stage it is useful to introduce a representation of the Dirac δ-functional:

$$\delta[\nabla \cdot \xi] = \int \mathscr{D}[\phi] \exp\left(i \int dx\, \phi(x)\, \nabla \cdot \xi(x) \right). \qquad (4.5.24)$$

Further, we make our first use of the notion of the generating functional for correlation functions. Consider a functional:

$$\mathscr{Z}[\mathbf{J}] = \mathscr{N}^{-1} \int \mathscr{D}[\xi]\, \mathscr{P}_R[\xi] \exp\left(\int dz\, \mathbf{J}(z) \cdot {}_{,}(z) \right), \qquad (4.5.25)$$

where $\mathscr{N}$ is the normalization factor. By differentiation the above expression with respect to the source fields $\mathbf{J}$ we obtain moments of the distribution:

$$\langle \xi_{i_1}(x_1) \ldots \xi_{i_n}(x_n) \rangle = \left. \frac{\delta^n \langle \exp\left(\int dz\, \mathbf{J}(z) \cdot \xi(z) \right) \rangle_R}{\delta J_{i_1}(x_1) \ldots \delta J_{i_n}(x_n)} \right|_{\mathbf{J}=\mathbf{0}}. \qquad (4.5.26)$$

The source fields are set to zero once all the necessary differentiation has been done.

Collecting all of the above results together we can specify the second moment in a form amendable to a direct computation:

$$\langle \xi_i(\mathbf{x}_1, t_1)\, \xi_j(\mathbf{x}_2, t_2) \rangle = \frac{\delta^2}{\delta J_i(x_1)\, \delta J_j(x_2)} \int \mathscr{D}[\xi]\, \mathscr{D}[\phi]\, \mathscr{P}[\xi]$$

$$\times \exp\left(\int dx\, [\mathbf{J}(x) + i\phi(x)\nabla] \cdot \xi(x) \right). \qquad (4.5.27)$$

The above are simply Gaussian functional integrals which we can compute. After some algebra we arrive at:

$$\left\langle \xi_i\left(\mathbf{x}_1, t_1\right) \xi_j\left(\mathbf{x}_2, t_2\right)\right\rangle = \left(\delta_{ij} - \nabla^{-2}\partial_i\partial_j\right) D\delta\left(\mathbf{x}_1 - \mathbf{x}_2\right) \delta\left(t_1 - t_2\right). \quad (4.5.28)$$

Under Elsasser transformations we can define new noise fields with modified amplitudes. Further, as a consequence of the transformation, the newly defined fields have a cross correlation. These new fields are:

$$\mathbf{f} = \xi + \psi, \quad \mathbf{g} = \xi - \psi. \quad (4.5.29)$$

Their statistics are:

$$\left\langle f_i\left(\mathbf{x}_1, t_1\right) f_j\left(\mathbf{x}_2, t_2\right)\right\rangle = 2\left(A + B\right) \mathrm{P}_{ij} D\left(\nabla\right) \delta\left(\mathbf{x}_1 - \mathbf{x}_2\right) \delta\left(t_1 - t_2\right), \quad (4.5.30)$$

$$\left\langle g_i\left(\mathbf{x}_1, t_1\right) g_j\left(\mathbf{x}_2, t_2\right)\right\rangle = 2\left(A + B\right) \mathrm{P}_{ij} D\left(\nabla\right) \delta\left(\mathbf{x}_1 - \mathbf{x}_2\right) \delta\left(t_1 - t_2\right), \quad (4.5.31)$$

$$\left\langle f_i\left(\mathbf{x}_1, t_1\right) g_j\left(\mathbf{x}_2, t_2\right)\right\rangle = 2\left(A - B\right) \mathrm{P}_{ij} D\left(\nabla\right) \delta\left(\mathbf{x}_1 - \mathbf{x}_2\right) \delta\left(t_1 - t_2\right), \quad (4.5.32)$$

where we have defined the projector operator as

$$\mathrm{P}_{ij} = \left(\delta_{ij} - \nabla^{-2}\partial_i\partial_j\right). \quad (4.5.33)$$

This insures that noise fields are now compatible with the incompressibility condition and Maxwell's equations. We can also define another operator which enters the dynamic equation:

$$\mathrm{R}_{ijk} = \partial_j \mathrm{P}_{ik}. \quad (4.5.34)$$

4.5.2 Functional Integral Formulation

We seek to re-write the MHD equations in terms of functional integrals. A detailed review on the properties of functional integrals can be found in [41]. Our analysis closely follows the technique described in [43], yet a more detailed analysis can be found in [42].

Let us introduce a compact notation for the equations of motion, namely, we write them symbolically as:

$$F_i^{(1)}\left[\mathbf{P}, \mathbf{Q}\right] = f_i, \quad i = 1, \ldots, d, \quad (4.5.35)$$

$$F_i^{(2)}\left[\mathbf{Q}, \mathbf{P}\right] = g_i, \quad i = 1, \ldots, d. \quad (4.5.36)$$

We make the assumption that the problem is well-defined, namely, that there exists a unique solution in terms of **P** and **Q** which satisfies the above equations. We denote the solution as $(\mathbf{P}_s, \mathbf{Q}_s)$. Now we make use of the following identity:

$$1 = \int \mathscr{D}[\mathbf{P}]\, \mathscr{D}[\mathbf{Q}]\, \delta[\mathbf{P} - \mathbf{P}_s]\, \delta[\mathbf{Q} - \mathbf{Q}_s]. \tag{4.5.37}$$

It is convenient to substitute the argument of the Dirac δ-functional by the corresponding equation of motion. Such a transformation introduces a Jacobian determinant:

$$1 = \int \mathscr{D}[\mathbf{P}]\, \mathscr{D}[\mathbf{Q}] \prod_{i=1}^{d} \delta\left[F_i^{(1)}[\mathbf{P}, \mathbf{Q}] - f_i\right] \delta\left[F_i^{(2)}[\mathbf{Q}, \mathbf{P}] - g_i\right] \mathscr{J}. \tag{4.5.38}$$

The specific form of the Jacobian determinant will be the subject of the next section. It will be shown that it can be dropped because it is field independent, so we drop it for the rest of our discussion. Let us now consider an arbitrary functional of the fields **P** and **Q**. We wish to compute some average properties of such a functional. Evidently, we have to take averages over different realizations of the noise:

$$\langle \mathrm{F} \rangle = \int \mathscr{D}[\mathbf{f}]\, \mathscr{D}\left[\mathbf{g}\right] \mathrm{F}[\mathbf{P}, \mathbf{Q}]\, \mathscr{P}\left[\mathbf{f}, \mathbf{g}\right], \tag{4.5.39}$$

where $\mathscr{P}$ is the functional probability distribution of the noise fields in Elsasser variables. We can now insert the identity Eq. (4.5.37) into the above equation:

$$\langle \mathrm{F} \rangle = \int \mathscr{D}[\mathbf{f}]\, \mathscr{D}\left[\mathbf{g}\right] \mathscr{D}[\mathbf{P}]\, \mathscr{D}[\mathbf{Q}]\, \mathrm{F}[\mathbf{P}, \mathbf{Q}]\, \mathscr{P}\left[\mathbf{f}, \mathbf{g}\right]$$
$$\prod_{i=1}^{d} \delta\left[F_i^{(1)}[\mathbf{P}, \mathbf{Q}] - f_i\right] \delta\left[F_i^{(2)}[\mathbf{Q}, \mathbf{P}] - g_i\right]. \tag{4.5.40}$$

At this point we can choose two different approaches. Both are equivalent and is a matter of personal preference. We can integrate out the noise fields, then the dynamics will be captured in $\mathscr{P}$. Such an approach would give a functional integral solely in terms of physical fields. For this reason it is sometimes referred to as the "minimalist" approach [43]. Alternatively we could introduce two auxiliary vector fields through the functional integral representation of the Dirac δ-functional. This is the method we choose to follow, partially because this formalism offers a slightly more convenient starting point for the RG calculation.

We have already come across functional representation of the Dirac δ-functional in the previous section, so we can write:

$$\begin{aligned}\langle \mathrm{F}\rangle = &\int \mathscr{D}[\mathbf{f}]\,\mathscr{D}[\mathbf{g}]\,\mathscr{D}[\mathbf{P}]\,\mathscr{D}[\mathbf{Q}]\,\mathrm{F}[\mathbf{P},\mathbf{Q}]\,\mathscr{P}[\mathbf{f},\mathbf{g}] \\ &\times \int \mathscr{D}[\mathbf{h}]\exp\left(i\int dx\left(F_i^{(1)}[\mathbf{P},\mathbf{Q}] - f_i(x)\right)h_i(x)\right) \\ &\times \int \mathscr{D}[\mathbf{n}]\exp\left(i\int dx\left(F_i^{(2)}[\mathbf{Q},\mathbf{P}] - g_i(x)\right)n_i(x)\right). \end{aligned} \tag{4.5.41}$$

Here we assume summation over repeated indicies. The summation over the indicies comes from the product of δ—functionals. The auxiliary fields are now coupled to the noise fields. The advantage of the above formulation is that the statistics of the noise is coupled to these new auxiliary fields rather than the equation of motion, like in the minimalist approach.

The functional probability distribution is specified as follows:

$$\begin{aligned}\mathscr{P}[\mathbf{f},\mathbf{g}] = &\exp\left[-\frac{1}{2}\int dxdy\left(\mathbf{f}(x)\cdot\Gamma_+^{-1}(x,y)\cdot\mathbf{f}(y) + \mathbf{g}(x)\cdot\Gamma_+^{-1}(x,y)\cdot\mathbf{g}(y)\right)\right] \\ &\times \exp\left[-\int dxdy\mathbf{g}(x)\cdot\Gamma_-^{-1}(x,y)\cdot\mathbf{f}(y)\right]. \end{aligned} \tag{4.5.42}$$

We only need to integrate over the real noise fields. The above is just a Gaussian integral so we obtain:

$$\begin{aligned}\langle \mathrm{F}\rangle = &\int \mathscr{D}[\mathbf{P}]\,\mathscr{D}[\mathbf{Q}]\,\mathscr{D}[\mathbf{h}]\,\mathscr{D}[\mathbf{n}]\,\mathrm{F}[\mathbf{P},\mathbf{Q}]\exp\left[-\int dxdy\mathbf{n}(x)\cdot\Gamma_-(x,y)\cdot\mathbf{h}(y)\right] \\ &\times \exp\left[-\frac{1}{2}\int dxdy\left(\mathbf{h}(x)\cdot\Gamma_+(x,y)\cdot\mathbf{h}(y) + \mathbf{n}(x)\cdot\Gamma_+(x,y)\cdot\mathbf{n}(y)\right)\right] \\ &\times \exp\left[i\int dxF_i^{(1)}[\mathbf{P},\mathbf{Q}]h_i(x) + i\int dxF_i^{(2)}[\mathbf{Q},\mathbf{P}]n_i(x)\right]. \end{aligned} \tag{4.5.43}$$

Finally, we can make a special choice of F such that it behaves as a generating functional. We have encountered the idea of the generating functional in the previous section, so here we simply state the form of F that we seek:

$$\mathrm{F}\left[\mathbf{j}^P,\mathbf{j}^Q,\mathbf{g}^h,\mathbf{g}^n\right] = \exp\left[\int \mathbf{j}^P\cdot\mathbf{P} + \int \mathbf{j}^Q\cdot\mathbf{Q} + \int \mathbf{g}^h\cdot\mathbf{h} + \int \mathbf{g}^n\cdot\mathbf{n}\right]. \tag{4.5.44}$$

We have suppressed the argument of the functions for compactness, but it is understood that the integrals run over the space-time arguments. This completes the specification of the functional integral for the MHD equations. By taking functional derivatives with respect to the source fields of F we obtain correlation functions of the theory.

4.5.3 *Jacobian of the Transformation*

4.5.3.1 Ordinary Calculus

To understand why we require a Jacobian upon a change of variables in the argument of the Dirac δ-functional we consider a simple example from ordinary calculus.

The Dirac δ-function can be defined in terms of a Gaussian function in the limit of the standard deviation approaching zero:

$$\delta(x) = \lim_{a \to 0} \delta_a(x), \tag{4.5.45}$$

where

$$\delta_a(x) = \frac{1}{a\sqrt{\pi}} \exp\left(-\frac{x^2}{a^2}\right). \tag{4.5.46}$$

Using this representation we can try to compute the integral:

$$I_1 = \int_{-\infty}^{\infty} dx \delta(f(x)), \quad f(x)|_{x=x_0} = 0. \tag{4.5.47}$$

We assume that the solution to $f(x) = 0$ is unique. We can generalize to the case of multiple solutions in a simple manner. For now we stick to a single solution case. A standard way of solving the integral is to change variables $x \to f$ and integrate over f. We would like to present an argument which is perhaps more in the spirit of field theory. Let us re-write the integral I using the Gaussian representation. We will eventually recover the limiting case by taking $a \to 0$:

$$I_{1,a} = \frac{1}{a\sqrt{\pi}} \int_{-\infty}^{\infty} dx \exp\left(-\frac{f(x)^2}{a^2}\right). \tag{4.5.48}$$

We assume that f is a well behaved analytic function and therefore it can be Taylor-expanded around x_0:

$$f(x)^2 = \left[(x - x_0) f'(x_0) + \frac{1}{2!}(x - x_0)^2 f''(x_0) + \cdots\right]^2. \tag{4.5.49}$$

The first term in the Taylor expansion is zero by definition. Let us make a change of variables:

$$y = x - x_0, \tag{4.5.50}$$

then

$$I_{1,a} = \frac{1}{a\sqrt{\pi}} \int_{-\infty}^{\infty} dy \exp\left(-\frac{1}{a^2} \left[yf'(x_0) + \frac{1}{2!} y^2 f''(x_0) + \cdots \right]^2 \right). \tag{4.5.51}$$

The expansion can be grouped into a quadratic part and the rest:

$$I_{1,a} = \int dx \frac{1}{a\sqrt{\pi}} \exp\left(-\frac{1}{a^2} y^2 f'(x_0)^2 \right) \exp\left(-\frac{1}{a^2} V(y) \right). \tag{4.5.52}$$

Importantly, $V(y)$ is of order y^3 and higher. Perform another transformation of variables:

$$t = \frac{y}{a}, \tag{4.5.53}$$

$$I_{1,a} = \int_{-\infty}^{\infty} dt \frac{1}{\sqrt{\pi}} \exp\left(-t^2 f'(x_0)^2 \right) \exp\left(-V'(t) \right). \tag{4.5.54}$$

All the dependence on the parameter a has shifted into $V'(t)$. However, as we have noted, terms in $V(y)$ are of order y^3 and higher. Hence, after the transformation $t = y/a$, the contribution $V'(t)$ is a polynomial in positive powers of a. When we take the limit $a \to 0$, all terms in $V'(t)$ go to zero. In field theory a similar technique exists. When one tries to perform a loop expansion, say in ϕ^4 theory, instead of a one has Planck's constant $\hbar$ [45]. It does not make sense to take the limit $\hbar \to 0$, instead it is used as a small parameter for a systematic expansion.

After taking the limit of $a \to 0$ we are left with a Gaussian integral:

$$I_1 = \lim_{a \to 0} I_{1,a} = \int dt \frac{1}{\sqrt{\pi}} \exp\left(-t^2 f'(x_0)^2 \right). \tag{4.5.55}$$

This is a straightforward integral to evaluate, so we read our final result:

$$\int_{-\infty}^{\infty} dx \delta(f(x)) = \frac{1}{f'(x_0)}. \tag{4.5.56}$$

The role of the Jacobian is thus to normalize the Dirac δ-function upon a change of variables:

$$\int_{-\infty}^{\infty} dx \delta(x - x_0) = \int_{-\infty}^{\infty} dx \delta(f(x)) f'(x). \tag{4.5.57}$$

The above considerations can be further generalized to multiple integrals. Let us consider the same problem but in $2D$. The argument of the Dirac δ-function is now a two-component vector function:

$$\mathbf{f}(\mathbf{x}) = (f_1(x, y), f_2(x, y)). \tag{4.5.58}$$

We want to solve the integral:

$$I_2 = \int_{-\infty}^{\infty} dxdy\delta(\mathbf{f}(\mathbf{x})), \quad \mathbf{f}(\mathbf{x})|_{(x,y)=(x_0,y_0)} = 0. \tag{4.5.59}$$

We assume that there is a unique solution to $\mathbf{f}(\mathbf{x}) = 0$ and it is an analytic function in both variables. By the vector argument of the Dirac δ-function it is understood that:

$$\delta(\mathbf{f}(\mathbf{x})) = \delta(f_1(x, y))\,\delta(f_2(x, y)). \tag{4.5.60}$$

Once again we make use of the Gaussian representation of the Dirac δ-function:

$$I_{2,a} = \frac{1}{a^2\pi}\int_{-\infty}^{\infty} dxdy \exp\left(-\frac{1}{a^2}\left(f_1(x, y)^2 + f_2(x, y)^2\right)\right). \tag{4.5.61}$$

Let us expand both components of $\mathbf{f}$ around the point (x_0, y_0):

$$f_1(x, y) = \partial_y f_1(y - y_0) + \partial_x f_1(x - x_0) + \cdots, \tag{4.5.62}$$
$$f_2(x, y) = \partial_y f_2(y - y_0) + \partial_x f_2(x - x_0) + \cdots. \tag{4.5.63}$$

The integral can be grouped into two contributions: one which is quadratic in variables x and y and the rest. Upon a shift of variables:

$$t = x - x_0, \quad z = y - y_0, \tag{4.5.64}$$

we can write:

$$I_{2,a} = \frac{1}{a^2\pi}\int_{-\infty}^{\infty} dtdz \exp\left(-\frac{1}{a^2}\left(\left[\partial_y f_1\right]^2 z^2 + [\partial_x f_1]^2 t^2 + 2\left[\partial_y f_1 \partial_x f_1\right] tz\right)\right)$$
$$\times \exp\left(-\frac{1}{a^2}\left(\left[\partial_y f_2\right]^2 z^2 + [\partial_x f_2]^2 t^2 + 2\left[\partial_y f_2 \partial_x f_2\right] tz\right) - \frac{1}{a^2}V(t, z)\right). \tag{4.5.65}$$

where the contribution $V(z,t)$ contains all the higher order terms. After rescaling the variables z and t by a and taking the limit $a \to 0$, only the quadratic part of the integral is non-vanishing. We are left with a $2D$ Gaussian integral:

$$I_2 = \frac{1}{\pi} \int_{-\infty}^{\infty} dt dz \exp\left(- \left(\left[\partial_y f_1 \right]^2 z^2 + \left[\partial_x f_1 \right]^2 t^2 + 2 \left[\partial_y f_1 \partial_x f_1 \right] tz \right) \right)$$
$$\times \exp\left(- \left(\left[\partial_y f_2 \right]^2 z^2 + \left[\partial_x f_2 \right]^2 t^2 + 2 \left[\partial_y f_2 \partial_x f_2 \right] tz \right) \right). \quad (4.5.66)$$

Further, the above expression can be more conveniently written in terms of matrices:

$$\mathbf{M}_i = \begin{pmatrix} \left[\partial_x f_i \right]^2 & \left[\partial_y f_i \right] \left[\partial_x f_i \right] \\ \left[\partial_y f_i \right] \left[\partial_x f_i \right] & \left[\partial_y f_i \right]^2 \end{pmatrix}, \quad i \in \{1, 2\}. \quad (4.5.67)$$

We define the following matrix:

$$\mathbf{M} = \mathbf{M}_1 + \mathbf{M}_2, \quad (4.5.68)$$

and we arrive at the expression:

$$\to I_2 = \frac{1}{\pi} \int_{-\infty}^{\infty} d\mathbf{x} \exp\left(-\mathbf{x}^T \cdot \mathbf{M} \cdot \mathbf{x} \right). \quad (4.5.69)$$

The matrix $\mathbf{M}$ is symmetric, so the last integral can be calculated by diagonalization. The result of the final integration reads:

$$I_2 = (\det \mathbf{M})^{-1/2}. \quad (4.5.70)$$

With a little bit of algebra it can be shown that:

$$(\det \mathbf{M})^{-1/2} = \frac{1}{\left[\partial_x f_1 \right] \left[\partial_y f_2 \right] - \left[\partial_x f_2 \right] \left[\partial_y f_1 \right]}. \quad (4.5.71)$$

The denominator is in fact the familiar Jacobian determinant. Thus, the value of the integral K is:

$$\int_{-\infty}^{\infty} dx dy \delta \left(\mathbf{f}(\mathbf{x}) \right) = \frac{1}{\mathcal{J}}, \quad \mathcal{J} = \det \begin{pmatrix} \partial_x f_1 & \partial_y f_1 \\ \partial_x f_2 & \partial_y f_2 \end{pmatrix}_{(x_0, y_0)}. \quad (4.5.72)$$

From the above example the generalization to arbitrary dimension follows, however the algebraic manipulations become more involved. We can state the general result:

$$\int d^d\mathbf{x}\delta\left(\mathbf{f}\left(\mathbf{x}\right)\right) = \int d^d\mathbf{x}\,\mathscr{J}^{-1}\delta\left(\mathbf{x}-\mathbf{x}_0\right). \tag{4.5.73}$$

A more general case of multiple solutions to the equation $\mathbf{f}\left(\mathbf{x}\right)|_{\mathbf{x}=\mathbf{x}_i} = 0$, requires making a Taylor expansion around every $\mathbf{x}_i$. The result is then a sum of Jacobian determinants evaluated at every $\mathbf{x}_i$.

4.5.3.2 Functional Jacobian

Often, it is convenient to think of functional integrals as ordinary integrals which are generalized to an infinite number of variables. In particular this is often the way one interprets Gaussian functional integrals.

$$\begin{aligned} K &= \int_{-\infty}^{\infty}\prod_{i=1}^{N} d\phi_i \exp\left(-\sum_{j,k}\phi_i A_{jk}\phi_k\right) \\ &\rightarrow \int \mathscr{D}\left[\phi\right]\exp\left(-\int d\mathbf{x}d\mathbf{y}\phi\left(\mathbf{x}\right)\mathbf{A}\left(\mathbf{x},\mathbf{y}\right)\phi\left(\mathbf{y}\right)\right). \end{aligned} \tag{4.5.74}$$

The normalization factor is often dropped. In the continuum limit it generates an infinite contribution. However, one is primarily concerned with average quantities so these infinities do not contribute.

In this respect the method of the Dirac δ-function representation through Gaussian integrals serves as a convenient tool for constructing functional Jacobian determinants. In particular, using the above analogy we can write:

$$\int \mathscr{D}\left[\phi\right]\delta\left[F\left[\phi\right]\right] = \int \mathscr{D}\left[\phi\right]\delta\left[\phi-\phi_s\right]\mathscr{J}^{-1}, \tag{4.5.75}$$

where the functional Jacobian determinant is

$$\mathscr{J} = \det\left(\frac{\delta F\left[\phi\left(\mathbf{x}\right)\right]}{\delta\phi\left(\mathbf{y}\right)}\right). \tag{4.5.76}$$

The meaning of the above expression is best understood if we go back to the discretized case, then the Jacobian matrix is specified by the following matrix elements:

$$J_{ij} = \frac{\partial F_i}{\partial\phi_j} \leftrightarrow J\left(\mathbf{x},\mathbf{y}\right) = \frac{\delta F\left[\phi\left(\mathbf{x}\right)\right]}{\delta\phi\left(\mathbf{y}\right)}. \tag{4.5.77}$$

Upon discretization the continuous variable $\phi\left(\mathbf{x}\right)$ is substituted by ϕ_i, where i denotes the lattice site of which suppose there are N. The constraint $F\left[\phi\right] = 0$ turns into N equations for each ϕ_i. Thus, the result turns into an integral with N variables, which can be calculated using the usual rules of calculus. A rigorous discussion on the points made above can be found in [42].

4.5.3.3 MHD Jacobian Determinant

With regards to the MHD equation in Elsasser variables, the Jacobian determinant is a generalization of the above ideas to functional integrals with two vector field variables, which is a determinant of a 2×2 block array:

$$\mathscr{J} = \det \begin{pmatrix} \left(\partial_t - \gamma_+ \nabla^2\right) \delta_{jk} + \mathrm{P}_{jk} Q_l \partial_l & -\gamma_- \nabla^2 \delta_{jk} + \mathrm{P}_{jn} \partial_k P_n \\ -\gamma_- \nabla^2 \delta_{jk} + \mathrm{P}_{jn} \partial_k Q_n & \left(\partial_t - \gamma_+ \nabla^2\right) + \mathrm{P}_{jk} P_l \partial_l \end{pmatrix} \delta\left(x - x'\right). \tag{4.5.78}$$

Formally, we can manipulate the Jacobian matrix by means of the identities:

$$\det \mathbf{A} = \exp\left(\mathrm{Tr} \ln \mathbf{A}\right) \quad \text{and} \quad \det\left(\mathbf{AB}\right) = \det\left(\mathbf{A}\right) \det\left(\mathbf{B}\right). \tag{4.5.79}$$

Let us define the Jacobian matrix as follows:

$$J = \partial_t \mathbf{1} + \mathbf{M}, \quad \mathbf{M} \equiv \begin{pmatrix} \frac{\delta F_{1,j}}{\delta P_k} & \frac{\delta F_{1,j}}{\delta Q_k} \\ \frac{\delta F_{2,j}}{\delta P_k} & \frac{\delta F_{2,j}}{\delta Q_k} \end{pmatrix}, \tag{4.5.80}$$

then:

$$\begin{aligned} \mathscr{J} &= \det\left(J\right) \\ &= \det\left(\partial_t \mathbf{1} + \mathbf{M}\right) \\ &= \det\left(\partial_t\right) \exp\left(\sum_{n=1}^{\infty} \frac{1}{n} \mathrm{Tr}\left(\partial_t^{-1} \mathbf{M}\right)^n\right). \end{aligned} \tag{4.5.81}$$

At this stage it is important to specify what we mean by ∂_t^{-1}. It is the inverse of the operator ∂_t and hence it can be viewed as a Green's function such that:

$$\partial_t G\left(t, t'\right) = \delta\left(t - t'\right), \tag{4.5.82}$$

with the Green's function being:

$$G\left(t, t'\right) = \Theta\left(t - t'\right). \tag{4.5.83}$$

To proceed we also need to clarify what we mean by trace, Tr, when dealing with functionals. In the context of the MHD it is:

$$\mathrm{Tr} \equiv \sum_{ij} \delta_{ij} \int d\mathbf{x} d\mathbf{y} dt dt' \delta\left(\mathbf{x} - \mathbf{y}\right) \delta\left(t - t'\right). \tag{4.5.84}$$

The sum runs over the diagonal element of the matrix $\mathbf{M}$ in the usual sense. Each diagonal element of $\mathbf{M}$ is a continuous representation of a discrete matrix in the sense of the Eq. (4.5.77). Thus, summing each diagonal element of these matricies in the continuum limit is represented by integration.

We can try to evaluate the trace part in the exponential term by term:

$$\begin{aligned}\operatorname{Tr}\left(G\left(t,t'\right)\mathbf{M}\right) &= \sum_{ij}\delta_{ij}\int d\mathbf{x}d\mathbf{y}dtdt'\delta\left(\mathbf{x}-\mathbf{y}\right)\delta\left(t-t'\right)\Theta\left(t-t'\right)\mathbf{M}_{ij}\left(x,y\right)\\ &= \Theta\left(0\right)\sum_{i}\int dtd\mathbf{x}d\mathbf{y}\delta\left(\mathbf{x}-\mathbf{y}\right)\mathbf{M}_{ii}\left(\mathbf{x},t,\mathbf{y},t\right)\\ &= \Theta\left(0\right)\sum_{i}\int dt\operatorname{Tr}_{\text{space}}\mathbf{M}_{ii}\left(\mathbf{x},t,\mathbf{y},t\right).\end{aligned} \tag{4.5.85}$$

The quantity $\Theta(0)$ is ill-defined. It is in fact specified by the particular description we choose when we put a continuous equation of motion on a lattice with finite time steps. This issue is discussed in more detail in [42, 46]. To put the above statement into context let us define the Heaviside step function as follows:

$$\Theta\left(x\right) = \int_{-\infty}^{x}\frac{d\Theta\left(y\right)}{dy}dy. \tag{4.5.86}$$

We now make use of the general definition of a derivative:

$$\Theta\left(x\right) = \int_{-\infty}^{x}\lim_{\delta y\to 0}\frac{\Theta\left(y+\frac{\alpha+1}{2}\delta y\right)-\Theta\left(y+\frac{\alpha-1}{2}\delta y\right)}{\delta y}dy. \tag{4.5.87}$$

Assume that $\delta y \to 0^{+}$, then:

$$\Theta\left(0\right) = \frac{\alpha+1}{2}. \tag{4.5.88}$$

It is clear from the above that the definition of the step function at zero depends on the discretization method.

The Heaviside step function appears in the equations only with the time arguments. Thus it is the way we discretized the time variable only, which dictates the choice of $\Theta(0)$. A convenient choice can be $\alpha = -1$, then the Jacobian determinant is simply an irrelevant constant which can be dropped. However, such an asymmetric choice of discretization breaks coordinate invariance in field space [43]. For this reason one usually sticks with a symmetric scheme, which leads to:

$$\Theta\left(0\right) = \frac{1}{2}. \tag{4.5.89}$$

It follows that the first term in the exponential, Eq. (4.5.81) can be written as:

$$\mathrm{Tr}\left(\partial_t^{-1}\mathbf{M}\right) = \frac{1}{2}\sum_i \int dt \mathrm{Tr}_{\mathrm{space}} \mathbf{M}_{ii}\left(\mathbf{x}, t, \mathbf{y}, t\right). \tag{4.5.90}$$

Conveniently, all other terms in Eq. (4.5.81) vanish. To see this we consider the second term in the exponential:

$$\begin{aligned}\mathrm{Tr}\left(G\left(t, t'\right)\mathbf{M}\right)^2 &= \int dt dt'' G\left(t, t''\right) G\left(t'', t\right) \int d\mathbf{x} d\mathbf{z} f\left(\mathbf{x}, t, \mathbf{z}, t''\right) \\ &= \int dt dt'' \Theta\left(t - t''\right) \Theta\left(t'' - t\right) \int d\mathbf{x} d\mathbf{z} f\left(\mathbf{x}, t, \mathbf{z}, t''\right) \\ &= 0. \end{aligned} \tag{4.5.91}$$

The function f contains all of the information about the matrix $\mathbf{M}^2$, but its exact form does not matter in the above argument. The product of Heaviside step functions kills the above contribution. Analogously one sees that higher order terms would exhibit the same property [43] and therefore we retain only the first term in the expansion in Eq. (4.5.81).

We can state the final form of the Jacobian determinant in the following form:

$$\mathcal{J} = \det\left(\partial_t\right) \exp\left(\frac{1}{2}\sum_i \int dt \mathrm{Tr}_{\mathrm{space}} \mathbf{M}_{ii}\left(\mathbf{x}, t, \mathbf{y}, t\right)\right). \tag{4.5.92}$$

Note, the prefactor of the exponential is field independent. As we have argued with the normalization of Gaussian functional integrals, the prefactor can be dropped from the formulation of the problem, since we are only interested in average quantities with respect to the functional probability distribution. For this reason the prefactor in Eq. (4.5.92) can also be dropped.

Finally we make the following observation about the matrix $\mathbf{M}$:

$$\mathbf{M}_{ij} = \mathcal{M}_{ij} \delta\left(\mathbf{x} - \mathbf{y}\right) \delta\left(t - t'\right), \tag{4.5.93}$$

which follows from a direct computation, Eq. (4.5.78). Then, we can write:

$$\begin{aligned}\mathcal{J} &= \exp\left(\frac{1}{2}\sum_i \int dt d\mathbf{x} d\mathbf{y} \delta\left(\mathbf{x} - \mathbf{y}\right) \mathcal{M}_{ii}\left(\mathbf{x}, t, \mathbf{y}, t\right) \delta\left(\mathbf{x} - \mathbf{y}\right) \delta\left(t - t'\right)\right) \\ &= \exp\left(d \int dt d\mathbf{x} d\mathbf{y} \delta\left(\mathbf{x} - \mathbf{y}\right) \left(-\gamma_+ \nabla^2\right) \delta\left(\mathbf{x} - \mathbf{y}\right) \delta\left(t - t'\right)\right) \\ &\times \exp\left(\frac{1}{2}\int dt d\mathbf{x} d\mathbf{y} \delta\left(\mathbf{x} - \mathbf{y}\right) \left(\mathrm{P}_{\mathrm{ik}} Q_l \partial_l + \mathrm{P}_{\mathrm{ik}} P_l \partial_l\right) \delta_{ki} \delta\left(\mathbf{x} - \mathbf{y}\right) \delta\left(t - t'\right)\right). \end{aligned} \tag{4.5.94}$$

The first part is field independent, so we can safely ignore it despite the presence of an ill-defined object such as $\delta(0)$ [44, 47]. The second contribution requires a little bit more work to be done with it. Sum over repeated indicies is assumed. Consider the quantity:

$$H = \int dt d\mathbf{x} d\mathbf{y} \delta(\mathbf{x} - \mathbf{y}) \mathrm{P}_{\mathrm{ik}} Q_l(\mathbf{x}, t) \partial_l \delta_{ki} \delta(\mathbf{x} - \mathbf{y}) \delta(t - t') . \tag{4.5.95}$$

The time-dependent delta-function can be integrated over. The projector operator when summed over the diagonal elements is:

$$\mathrm{P}_{\mathrm{ik}} \delta_{ki} = d - 1. \tag{4.5.96}$$

Therefore we have:

$$H = (d-1) \int d\mathbf{y} \left(\int d\mathbf{x} \delta(\mathbf{x} - \mathbf{y}) Q_l(\mathbf{x}, t') \partial_l \delta(\mathbf{x} - \mathbf{y}) \right) . \tag{4.5.97}$$

The integral over $\mathbf{x}$ can be performed by parts:

$$\int d\mathbf{x} \left[\delta(\mathbf{x} - \mathbf{y}) Q_l(\mathbf{x}, t') \right] \partial_l \delta(\mathbf{x} - \mathbf{y}) = - \int d\mathbf{x} \delta(\mathbf{x} - \mathbf{y}) \partial_l \left[\delta(\mathbf{x} - \mathbf{y}) Q_l(\mathbf{x}, t') \right] . \tag{4.5.98}$$

Note, because $\nabla \cdot \mathbf{Q} = 0$, it follows:

$$\int d\mathbf{x} \delta(\mathbf{x} - \mathbf{y}) Q_l(\mathbf{x}, t') \partial_l \delta(\mathbf{x} - \mathbf{y}) = - \int d\mathbf{x} \delta(\mathbf{x} - \mathbf{y}) Q_l(\mathbf{x}, t') \partial_l \delta(\mathbf{x} - \mathbf{y}) . \tag{4.5.99}$$

The surface term vanishes because of the condition on the fields we have imposed earlier. We conclude that this contribution to the Jacobian is zero since the relationship of the form $A = -A$ implies $A = 0$. From the above we can see that the Jacobian determinant is field independent, so we can ignore this diverging contribution in the same manner as we have ignored infinite normalization in functional Gaussian integrals. For the remaining discussion we simply drop the Jacobian determinant from our formulation.

4.6 Free Theory

The functional integral formulation of the MHD equations can be understood solely in terms of the Gaussian integrals. In fact, a Gaussian theory forms the starting point for any perturbative analysis of the problem. To see how this comes about, let us split the equations of motion into linear and non-linear components:

$$F_i^{(1)}[\mathbf{P},\mathbf{Q}] = F_{i,\mathrm{L}}^{(1)}[\mathbf{P},\mathbf{Q}] + F_{i,\mathrm{Int}}^{(1)}[\mathbf{P},\mathbf{Q}], \tag{4.6.1}$$

$$F_i^{(2)}[\mathbf{Q},\mathbf{P}] = F_{i,\mathrm{L}}^{(2)}[\mathbf{Q},\mathbf{P}] + F_{i,\mathrm{Int}}^{(1)}[\mathbf{Q},\mathbf{P}]. \tag{4.6.2}$$

We can split the generating functional accordingly:

$$\begin{aligned}
\mathscr{Z} = & \int \mathscr{D}[\mathbf{P}]\mathscr{D}[\mathbf{Q}]\mathscr{D}[\mathbf{h}]\mathscr{D}[\mathbf{n}] \exp\left[-\int dxdy \mathbf{n}(x)\cdot\Gamma_-(x,y)\cdot\mathbf{h}(y)\right] \\
& \times \exp\left[-\frac{1}{2}\int dxdy\left(\mathbf{h}(x)\cdot\Gamma_+(x,y)\cdot\mathbf{h}(y) + \mathbf{n}(x)\cdot\Gamma_+(x,y)\cdot\mathbf{n}(y)\right)\right] \\
& \times \exp\left[i\int dx F_{i,\mathrm{L}}^{(1)}[\mathbf{P},\mathbf{Q}]h_i(x) + i\int dx F_{i,\mathrm{L}}^{(2)}[\mathbf{Q},\mathbf{P}]n_i(x)\right] \\
& \times \exp\left[\int dx \mathbf{j}^P(x)\cdot\mathbf{P}(x) + \int dx\mathbf{j}^Q(x)\cdot\mathbf{Q}(x) + \int dx\mathbf{g}^h(x)\cdot\mathbf{h}(x) + \int dx\mathbf{g}^n(x)\cdot\mathbf{n}(x)\right] \\
& \times \exp\left[i\int dx F_{i,\mathrm{Int}}^{(1)}[\mathbf{P},\mathbf{Q}]h_i(x) + i\int dx F_{i,\mathrm{Int}}^{(1)}[\mathbf{Q},\mathbf{P}]n_i(x)\right].
\end{aligned} \tag{4.6.3}$$

Let us write the interaction term as a series expansion of the exponential. For book-keeping purposes we introduce a parameter λ, which labels the non-linear terms:

$$\mathscr{Z} = \mathscr{Z}_0 \sum_{n=0}^{\infty} i^n\lambda^n \left(\int dx F_{i,\mathrm{Int}}^{(1)}[\mathbf{P},\mathbf{Q}]h_i(x) + \int dx F_{i,\mathrm{Int}}^{(1)}[\mathbf{Q},\mathbf{P}]n_i(x)\right)^n. \tag{4.6.4}$$

The prefactor $\mathscr{Z}_0$ stands for the part of the functional integral which is quadratic in the field. The series expansion forms a polynomial in the fields. These in turn can be expressed as derivatives with respect to the source fields. Thus we can write:

$$\begin{aligned}
\mathscr{Z} = & \sum_{n=0}^{\infty} i^n\lambda^n \left(\int dx F_{i,\mathrm{Int}}^{(1)}\left[\frac{\delta}{\delta\mathbf{j}^P}, \frac{\delta}{\delta\mathbf{j}^Q}\right]\frac{\delta}{\delta g_i^h(x)}\right. \\
& \left. + \int dx F_{i,\mathrm{Int}}^{(1)}\left[\frac{\delta}{\delta\mathbf{j}^Q}, \frac{\delta}{\delta\mathbf{j}^P}\right]\frac{\delta}{\delta g_i^n(x)}\right) \mathscr{Z}_0\left[\mathbf{j}^P, \mathbf{j}^Q, \mathbf{g}^h, \mathbf{g}^n\right].
\end{aligned} \tag{4.6.5}$$

Thus, the generating functional of the linear theory forms the central object from which we can calculate the correlation function of the complete theory to the desired accuracy.

The generating functional of a free theory consists solely of Gaussian integrals, so we can integrate out all the fields. At this point it is convenient to work in the Fourier representation of the MHD equations. Furthermore, since our goal is to perform a Wilsonian RG calculation we are obliged to work in Fourier space.

The equations of motion can be conveniently written as follows:

$$G_0^{-1}(k)\begin{pmatrix} P_l(k) \\ Q_l(k)\end{pmatrix} = \begin{pmatrix} f_l(k) \\ g_l(k)\end{pmatrix} - \lambda i \mathrm{R}_{lik}(\mathbf{k})\begin{pmatrix}\int dq Q_i(q)P_k(k-q) \\ \int dq P_i(q) Q_k(k-q)\end{pmatrix}, \tag{4.6.6}$$

where we have defined:

$$G_0^{-1}(k) = \begin{pmatrix} i\omega + \gamma_+ k^2 & \gamma_- k^2 \\ \gamma_- k^2 & i\omega + \gamma_+ k^2 \end{pmatrix}, \tag{4.6.7}$$

$$\mathrm{R}_{lik}(\mathbf{k}) = k_i \left(\delta_{lk} - \frac{k_l k_k}{k^2} \right). \tag{4.6.8}$$

To make the equation more tractable we introduce further notation, using higher dimensional vectors:

$$\mathbf{R} = \begin{pmatrix} \mathbf{P} \\ \mathbf{Q} \end{pmatrix}, \quad \mathbf{H} = \begin{pmatrix} \mathbf{h} \\ \mathbf{n} \end{pmatrix}, \quad \mathbf{J} = \begin{pmatrix} \mathbf{j}^P \\ \mathbf{j}^Q \end{pmatrix}, \quad \mathbf{G} = \begin{pmatrix} \mathbf{g}^h \\ \mathbf{g}^n \end{pmatrix}, \quad \mathbf{M} = \begin{pmatrix} \Gamma_+ & \Gamma_- \\ \Gamma_- & \Gamma_+ \end{pmatrix}. \tag{4.6.9}$$

Using the above notation the generating functional can be written as:

$$\mathscr{Z}_0 = \int \mathscr{D}[\mathbf{R}]\,\mathscr{D}[\mathbf{H}] \exp\left[-\frac{1}{2} \int dkdq\,\mathbf{H}(k)\,\mathbf{M}(k,q)\,\mathbf{H}(q) + \int dk\mathbf{G}(-k)\,\mathbf{H}(k) \right]$$
$$\times \exp\left[i \int dk\mathbf{H}^T(-k)\,G_0^{-1}(k)\,\mathbf{R}(k) + \int dk\mathbf{J}(-k)\,\mathbf{R}(k) \right]. \tag{4.6.10}$$

In this notation the Gaussian integral is more transparent. By performing the above integrals we obtain:

$$\mathscr{Z}_0 = \exp\left(\frac{1}{2} \int dk\mathbf{J}(-k)\,G_0(-k)\,\mathbf{M}(k)\,G_0(k)\,\mathbf{J}(k) \right)$$
$$\times \exp\left(i \int dk\mathbf{G}(-k)\,G_0(k)\,\mathbf{J}(k) \right). \tag{4.6.11}$$

We recover the correlation functions by differentiating $\mathscr{Z}_0$ with respect to corresponding source field:

$$\langle ih_i(p)\,P_j(q)\rangle = \langle in_i(p)\,Q_j(q)\rangle = C_{ij}^+(p)\,\delta(p+q), \tag{4.6.12}$$

$$\langle in_i(p)\,P_j(q)\rangle = \langle ih_i(p)\,Q_j(q)\rangle = C_{ij}^-(p)\,\delta(p+q), \tag{4.6.13}$$

where the propagators are:

$$C_{ij}^+(p) = \frac{1}{2}\left(\frac{1}{i\omega + \nu p^2} + \frac{1}{i\omega + \mu p^2} \right) \delta_{ij}, \tag{4.6.14}$$

$$C_{ij}^-(p) = \frac{1}{2}\left(\frac{1}{i\omega + \nu p^2} - \frac{1}{i\omega + \mu p^2} \right) \delta_{ij}. \tag{4.6.15}$$

and

$$\left\langle P_i(p) P_j(q)\right\rangle = \left\langle Q_i(p) Q_j(q)\right\rangle = G^+_{ij}(p)\,\delta(p+q), \tag{4.6.16}$$

$$G^+_{ij}(p) = 2\,|\mathbf{p}|^{-y}\,\mathrm{P}_{ij}\left(\frac{A}{\omega^2+\nu^2 p^4} + \frac{B}{\omega^2+\mu^2 p^4}\right), \tag{4.6.17}$$

$$\left\langle P_i(p) Q_j(q)\right\rangle = \left\langle Q_i(p) P_j(q)\right\rangle = G^-_{ij}(p)\,\delta(p+q), \tag{4.6.18}$$

$$G^-_{ij}(p) = 2\,|\mathbf{p}|^{-y}\,\mathrm{P}_{ij}\left(\frac{A}{\omega^2+\nu^2 p^4} - \frac{B}{\omega^2+\mu^2 p^4}\right). \tag{4.6.19}$$

These propagators are all we need to know. By means of Wick's theorem [47], all the non-linear terms in the series expansion of the interacting part of the action can be represented as combinations of these four propagators. The factor of 2 in the physical fields propagators is a commonly used convention, which we adopt for later comparison of our results with the literature.

Note, the propagators of the form $\left\langle Q_i(p) Q_j(q)\right\rangle$ can be calculated directly from the dynamic equations which is in fact a much simpler calculation. On the contrary, the propagators which involve auxiliary fields can be directly read-off in the functional integral formulation and are non-trivial if one is to compute them from the equations of motion.

4.7 Renormalization Group Transformations

4.7.1 Coarse Graining

The first step in the RG program is coarse graining, where we systematically remove small scale fluctuations. It is natural to introduce the smallest scale into the system; call it Λ. The fluid description has a minimum length beyond which it breaks down. For example, in turbulent flows such a scale is given by the viscosity. All of the relevant physics is captured between this dissipation scale and the macro-scale. Thus, we can think of Λ as the scale at which the energy is dissipated in the system by viscosity. Note, we refer to Λ as a length scale in Fourier space, so the corresponding physical length is of order Λ^{-1}.

By separating the small scale fluctuations we effectively separate out the Fourier modes of order Λ:

$$[0, \Lambda] \to \left[0, \Lambda b^{-1}\right] \bigcup \left[\Lambda b^{-1}, \Lambda\right], \tag{4.7.1}$$

where b is a parameter such that $b > 1$. Due to the presence of non-linear terms in the action only the linear term can be factorized upon such mode splitting:

$$\mathscr{Z}_0 = \left(\int \mathscr{D}[\mathbf{R}_<]\,\mathscr{D}[\mathbf{H}_<]\,e^{\mathscr{A}_0^<}\right)\left(\int \mathscr{D}[\mathbf{R}_>]\,\mathscr{D}[\mathbf{H}_>]\,e^{\mathscr{A}_0^>}\right). \tag{4.7.2}$$

Let us symbolically write the non-linear part of the action as follows:

$$\mathscr{A}_I\left[\mathbf{R},\mathbf{H}\right]=i\lambda\int dx\left(F^{(1)}_{i,\mathrm{Int}}\left[\mathbf{P},\mathbf{Q}\right]h_i\left(x\right)+F^{(1)}_{i,\mathrm{Int}}\left[\mathbf{Q},\mathbf{P}\right]n_i\left(x\right)\right). \tag{4.7.3}$$

Thus, we can write:

$$\mathscr{Z}=\int\mathscr{D}\left[\mathbf{R}_<\right]\mathscr{D}\left[\mathbf{H}_<\right]e^{\mathscr{A}_0^<}\left(\int\mathscr{D}\left[\mathbf{R}_>\right]\mathscr{D}\left[\mathbf{H}_>\right]e^{\mathscr{A}_0^>+\mathscr{A}_I[\mathbf{R},\mathbf{H}]}\right). \tag{4.7.4}$$

The expression cannot be evaluated exactly. However we can make use of the notion of the functional average with respect to the fast modes free theory:

$$\mathscr{Z}=\mathscr{Z}_0^>\int\mathscr{D}\left[\mathbf{R}_<\right]\mathscr{D}\left[\mathbf{H}_<\right]e^{\mathscr{A}_0^<}\left\langle e^{\mathscr{A}_I[\mathbf{R},\mathbf{H}]}\right\rangle_>, \tag{4.7.5}$$

where we have defined the functional average as:

$$\left\langle e^{\mathscr{A}_I[\mathbf{R},\mathbf{H}]}\right\rangle_>=\frac{\int\mathscr{D}\left[\mathbf{R}_>\right]\mathscr{D}\left[\mathbf{H}_>\right]e^{\mathscr{A}_0^>+\mathscr{A}_I[\mathbf{R},\mathbf{H}]}}{\int\mathscr{D}\left[\mathbf{R}_>\right]\mathscr{D}\left[\mathbf{H}_>\right]e^{\mathscr{A}_0^>}}. \tag{4.7.6}$$

This can be written in yet another way by exponentiating the average:

$$\mathscr{Z}=\mathscr{Z}_0^>\int\mathscr{D}\left[\mathbf{R}_<\right]\mathscr{D}\left[\mathbf{H}_<\right]\exp\left(\mathscr{A}_0^<+\ln\left\langle\exp\left(\mathscr{A}_I\left[\mathbf{R},\mathbf{H}\right]\right)\right\rangle_>\right). \tag{4.7.7}$$

The above expression is the central equation from which we proceed and form the cumulant expansion. Note that formally we end up with a partition function which describes a theory on the scale $\left[0,\Lambda b^{-1}\right]$. To manage the calculation in a more compact manner we employ diagrammatic techniques.

In order to proceed we need to define some diagrammatic rules. There are two types of non-linear terms which we can identify:

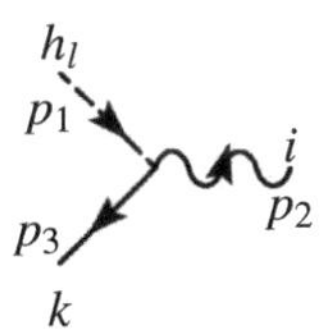

$$= \lambda \mathrm{R}_{lik}(-p_1)\delta(p_1+p_2+p_3)$$

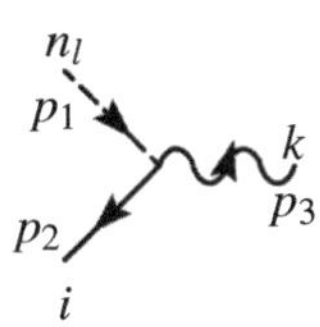

$$= \lambda \mathrm{R}_{lik}(-p_1)\delta(p_1+p_2+p_3)$$

Note, the difference between the two is the order in which physical fields are contracted. We need to keep track of such indices's throughout the calculation. Further, we have identified four types of free propagators in our theory. For each, we also write a special diagrammatic symbol:

$C^{\pm}_{ji}(p) =$  , $G^{\pm}_{ji}(p) =$  (4.7.8)

Using the above representation, we are now in a position to calculate corrections to the free theory from fast modes elimination by means of combining and permuting the diagrams [48]. The detail of the analysis can be found in the Appendix **B**.

4.7.2 Rescaling

At this stage we also need to address a question of how various quantities scale under spatial and temporal transformations. In contrast to the usual RG scaling procedure we have used in the Coulomb gas problem, in the dynamic RG we have to introduce an additional temporal exponent. Thus, the scaling transformations is defined as follows:

$$\mathbf{p}' = b\mathbf{p}, \qquad \omega' = b^z\omega. \tag{4.7.9}$$

Then, we can write for the fields:

$$P\left(b^{-1}\mathbf{p}', b^{-z}\omega'\right) = b^{\eta}P'\left(\mathbf{p}', \omega'\right), \qquad Q\left(b^{-1}\mathbf{p}', b^{-z}\omega'\right) = b^{\eta}Q'\left(\mathbf{p}', \omega'\right), \tag{4.7.10}$$

$$h\left(b^{-1}\mathbf{p}', b^{-z}\omega'\right) = b^{\xi}h'\left(\mathbf{p}', \omega'\right), \qquad n\left(b^{-1}\mathbf{p}', b^{-z}\omega'\right) = b^{\xi}n'\left(\mathbf{p}', \omega'\right). \tag{4.7.11}$$

We determine the scaling behavior of the coefficients by demanding that equations of motion remain invariant under such transformation. For example, the linear part of the action then transforms as

$$b^{-d-z}\int d\mathbf{p}d\omega b^{\xi}\mathbf{h}\left(-p\right)\left(ib^{-z}\omega + \gamma_{+}b^{-2}p^{2}\right)b^{\eta}\mathbf{P}\left(p\right). \tag{4.7.12}$$

We demand that the first order time derivative of the action remains invariant. This is true of any dynamical system governed by a first order time derivative since we can always rescale this term such that its coefficient remains at one. Thus we can write:

$$\xi + \eta = d + 2z. \tag{4.7.13}$$

This allows to write for the coefficients:

$$\gamma'_{\pm} = b^{z-2}\gamma_{\pm}. \tag{4.7.14}$$

Similarly we can derive the scaling relations for λ:

$$\lambda' = b^{z+\eta-1}\lambda. \tag{4.7.15}$$

These considerations become handy when we aim to restore the cut-off of the large-scale theory from Λb^{-1} to Λ. The detailed examination of this step can be found in the Appendix **E**. This step completes the series of steps involved in the RG program.

4.7.3 Differential RG Equations

After calculating explicitly all the loop integrals together with differential scale transformations we arrive at the following set of equations:

$$\frac{d\nu}{d\ell} = \left(z - 2 + \frac{\lambda^2}{\nu}\left[\frac{A_0}{\nu^2}\left(\frac{d^2-y-4}{2d\,(d+2)}\right) + \frac{B_0}{\mu^2}\left(\frac{y+d^2}{2d\,(d+2)}\right)\right.\right.$$
$$\left.\left. + \frac{(d^2-2)}{2d\,(d+2)}\left(\frac{A_2}{\nu^2} + \frac{B_2}{\mu^2}\right)\Lambda^{d-2}\right]\frac{S_d}{(2\pi)^d}\Lambda^{\epsilon}\right)\nu, \tag{4.7.16}$$

$$\frac{d\mu}{d\ell} = \left(z - 2 + \frac{\lambda^2\Lambda^{\epsilon}}{\mu\,(\nu+\mu)}\left[\frac{A}{\nu}\frac{(d-1)}{d} + \frac{B}{\mu}\frac{(d-3)}{d}\right]\frac{S_d}{(2\pi)^d}\right)\mu, \tag{4.7.17}$$

$$\frac{d\lambda}{d\ell} = \left(\eta - d - 1 - \frac{\lambda^2\Lambda^{\epsilon}}{d\,(d+2)}\frac{S_d}{(2\pi)^d}A\left[\frac{1}{\mu\nu\,(\nu+\mu)} + \frac{\mu}{\nu^2\,(\nu+\mu)^2}\right]\right.$$
$$\left. + \frac{\lambda^2\Lambda^{\epsilon}}{d\,(d+2)}\frac{S_d}{(2\pi)^d}B\left[\frac{1}{2\mu^3} + \frac{1}{\mu\nu\,(\nu+\mu)} + \frac{\nu}{\mu^2\,(\nu+\mu)^2}\right]\right)\lambda, \tag{4.7.18}$$

$$\frac{dA_0}{d\ell} = (-d - z + 2\xi + y)\,A_0, \tag{4.7.19}$$

$$\frac{dB_0}{d\ell} = (-d - z + 2\xi + y)\,B_0, \tag{4.7.20}$$

$$\frac{dA_2}{d\ell} = \left(-d - z + 2\xi - 2 + \frac{\lambda^2}{A_2}\frac{d^2-2}{d\,(d+2)}\left[\frac{A\,(\Lambda)^2}{\nu^3} + \frac{B\,(\Lambda)^2}{\mu^3}\right]\frac{S_d}{(2\pi)^d}\Lambda^{\tilde{\epsilon}}\right)A_2, \tag{4.7.21}$$

$$\frac{dB_2}{d\ell} = \left(-d - z + 2\xi - 2 + \frac{\lambda^2}{B_2}\frac{d-2}{d}\left[\frac{2A\,(\Lambda)\,B\,(\Lambda)}{\nu\mu\,(\nu+\mu)}\right]\frac{S_d}{(2\pi)^d}\Lambda^{\tilde{\epsilon}}\right)B_2, \tag{4.7.22}$$

where we have used the following definitions:

$$\epsilon = d - y - 4, \quad \tilde{\epsilon} = d - 2y - 6. \tag{4.7.23}$$

These equations form the main result of this part of the thesis. The equations describe the behaviour of the coefficients under an infinitesimal change of spatial and temporal scales together and an infinitesimal reduction of degrees of freedom. While some parts of these equations can be reduced to the results that can be found in the literature, which will be the subject of the next section, the equation for λ is new. In our calculation the correction to λ is of order λ^2. Some methods, such as DIA, do not go to that order [20], others simply stop at λ^2 [36], since the correction is a result of including all 1-loop corrections. Therefore there is no way of checking this result with previous studies. In what follows, we will focus on the regimes where such corrections are not important however further investigation into the nature of such higher order corrections is required and calls for a separate study. This will be touched upon in the upcoming sections.

4.7.4 Consistency Check

The differential RG equations look rather complicated, so it is useful to invoke at least a partial consistency check. For this reason we consider a particular regime, namely we restrict the exponent y to simplify the above equations. Note, the choice of $y \geq -2$ makes the coefficients A_2 and B_2 irrelevant near a fixed point i.e. the RG flow along these variables always points towards a fixed point. So in this regime we can ignore noise corrections.

Further, let us now consider how we would reduce our original coupled set of equations to pure NS equation. Simply by looking at the MHD equations we could set $\mathbf{b} = \mathbf{0}$, but by looking solely at the transformation of coefficients we cannot do that. Instead, we can try to take the limit $\mu \to \infty$. What does this limit mean for the solution $\mathbf{b}$? Consider first a regime $\mu \gg 1$, which in effect means that the magnetic Reynolds number is very small, $\mathscr{R}_m \ll 1$. The role of the Reynolds number is to measure the competition between linear and non-linear terms. In our regime, the induction equation is dominated by the linear term in the above case. The equation now has the form of a diffusion equation, which in dimensionless form is:

$$\frac{\partial \tilde{\mathbf{B}}}{\partial t} = \frac{1}{\mathscr{R}_m} \nabla^2 \tilde{\mathbf{B}} + \Psi. \tag{4.7.24}$$

We can switch off the noise by setting the amplitude of the noise covariance to zero, $B = 0$. Then we are left with:

$$\frac{\partial \tilde{\mathbf{B}}}{\partial t} = \frac{1}{\mathscr{R}_m} \nabla^2 \tilde{\mathbf{B}}. \tag{4.7.25}$$

The fundamental solution of this equation can be computed by Fourier transforms. More importantly, the quantity $\mathscr{R}_m^{-1}$, plays the role of the diffusion constant. Finally if we set $\mu \to \infty$, then $\mathscr{R}_m \to 0$ and $\mathscr{R}_m^{-1} \to \infty$, which in turn sets the solution $\tilde{\mathbf{B}} \to 0$.

Taking the above effects into account we conclude that the RG equations for the pure noisy NS equation are:

$$\frac{d\nu}{d\ell} = \left(z - 2 + \frac{A\lambda^2}{\nu^3} \left(\frac{d^2 - y - 4}{2d\,(d+2)} \right) \frac{S_d}{(2\pi)^d} \Lambda^{\epsilon} \right) \nu, \tag{4.7.26}$$

$$\frac{d\lambda}{d\ell} = (\eta - d - 1)\,\lambda, \tag{4.7.27}$$

$$\frac{dA}{d\ell} = (-d - z + 2\xi + y)\,A. \tag{4.7.28}$$

This set of equations is identical to that obtained previously [24].

References

1. M. Parviz, K. John, Tackling turbulence with supercomputers. Sci. Am. **276**(1), 62 (1997)
2. E.R. Priest, A.W. Hood, *Advances in Solar System Magnetohydrodynamics* (Cambridge University Press, Cambridge, 1991)
3. T. Ogino, R.J. Walker, M. Ashour-Abdalla, A three-dimensional MHD simulation of the interaction of the solar wind with Comet Halley. J. Geophys. Res. **93**(A9), 9568 (1988)
4. K.G. Powell, A.F. Nagy, Global MHD simulations of space plasma environments: heliosphere, comets, magnetospheres of planets and satellites. Astrophys. Space Sci. **274**, 407 (2000)
5. R.D. Simitev, F.H. Busse, A.G. Kosovichev, *Turbulent 3D MHD Dynamo Model in Spherical Shells: Regular Oscillations of the Dipolar Field*, eprint arXiv/1102.1092 (2011)
6. P.A. Durbin, B.A. Pettersson Reif, *Statistical Theory and Modelling of Turbulent Flows* (Wiley, Chichester, 2011)
7. W.D. McComb, *The Physics of Fluid Turbulence* (Oxford University Press, Oxford, 1990)
8. U. Frisch, *Turbulence* (Cambridge University Press, Cambridge, 1995)
9. D. Ruelle, *Chaotic Evolution and Strange Attractors* (Cambridge University Press, Cambridge, 1989)
10. M.J. Vishik, A.V. Fursikov, *Mathematical Problems of Statistical Hydrodynamics* (Kluwer, Dordrecht, 1988)
11. A.N. Kolmogorov, The local structure of turbulence in incompressible viscous fluid for very large Reynolds number. Dokl. Akad. Nauk SSSR **30**, 9 (1941)
12. A.N. Kolmogorov, On degeneration of isotropic turbulence in an incompressible viscous liquid. Dokl. Akad. Nauk SSSR **31**, 538 (1941)
13. A.N. Kolmogorov, Dissipation of energy in locally isotropic turbulence. Dokl. Akad. Nauk SSSR **32**, 16 (1941)
14. A.N. Kolmogorov, On logarithmically normal law of distribution on the size of particles under pulverization. Dokl. Akad. Nauk SSSR **31**, 99 (1941)
15. E. Hopf, J. Ration. Mech. Anal. **1**, 87 (1952)
16. H.H. Shen, A.A. Wray, Stationary turbulent closure via the hopf functional equation. J. Stat. Phys. **62**, 33 (1990)
17. R.M. Lewis, R.H. Kraichnan, A space-time functional formalism for turbulence. Commun. Pure Appl. Math. **15**, 397 (1962)
18. R.H. Kraichnan, The structure of isotropic turbulence at very high Reynolds numbers. J. Fluid Mech. **5**, 497 (1959)
19. G.F. Mazenko, *Equilibrium Statistical Mechanics* (Wiley, New York, 2000)
20. Y. Zhou, Renormalization group theory for fluids and plasma turbulence. Phys. Rep. **488**(1), 1–49 (2010)
21. R.H. Kraichnan, Hydrodynamic turbulence and the renormalization group theory. Phys. Rev. A **25**, 3281 (1982)
22. A. Einstein, Über die von der molekularkinetischen Theorie der Wärme geforderte Bewegung von in ruhenden Flüssigkeiten suspendierten Teilchen. Ann Phys **17**, 549 (1905)

23. S. Ma, G.F. Mazenko, Critical dynamics of ferromagnets in 6-ϵ dimensions: general discussion and detailed calculation. Phys. Rev. B **11**, 4077 (1975)
24. D. Forster, D.R. Nelson, M.J. Stephen, Large distance and long time properties of a randomly stirred fluid. Phys. Rev. A **16**, 732 (1977)
25. C. de Dominicis, P.C. Martin, Energy spectra of certain randomly-stirred fluids. Phys. Rev. A **19**, 419 (1979)
26. J.D. Fournier, U. Frisch, Remarks on the renormalization group in statistical fluid mechanics. Phys. Rev. A **28**, 1000 (1983)
27. V. Yakhot, S.A. Orszag, Renormalization group analysis of turbulence. Phys. Rev. Lett. **57**, 1722 (1986)
28. G.L. Eyink, The renormalization group method in statistical hydrodynamics. Phys. Fluids **6**, 3063 (1994)
29. L.M. Smith, W.C. Reynolds, On the Yakhot-Orszag renormalization group methods for deriving turbulence statistics and models. Phys. Fluids A **4**, 364 (1992)
30. X.H. Wang, F. Wu, One modification to the Yakhot-Orszag calculation in the renormalization-group theory of turbulence. Phys. Rev. E **48**, R37 (1993)
31. M.K. Nandy, Symmetrization of the self-energy integral in the Yakhot-Orszag renormalization-group calculation. Phys. Rev. E **55**, 5455 (1997)
32. J.-D. Fournier, P.-L. Sulem, A. Pouquet, Infrared properties if forced magnetohydrodynamic turbulence. J. Phys. A: Math. Gen. **15**, 1393 (1982)
33. W.M. Elsässer, The hydromagnetic equations. Phys. Rev. **79**, 183 (1950)
34. W.D. McComb, Galilean invariance and vertex renormalization in turbulence theory. Phys. Rev. E **71**, 037301 (2005)
35. A. Berera, D. Hochberg, Galilean invariance and homogeneous anisotropic randomly stirred flows. Phys. Rev. E **72**, 057301 (2005)
36. S.J. Camargo, H. Tasso, Renormalization group in magnetohydrodynamics turbulence. Phys. Fluids B **4**(5), 1199–1212 (1992)
37. A. Berera, D. Hochberg, *Asymptotic Properties of Turbulent Magnetohydrodynamics*, arXiv/cond-mat/0103447 (2006)
38. A. Berera, D. Hochberg, Gauge symmetry and Slavnov-Taylor identities for randomly stirred fluids. Phys. Rev. Lett. **99**, 254501 (2007)
39. C.-B. Kim, T.-J. Yang, Renormalization-group analysis on the scaling of the randomly-stirred magnetohydrodynamic plasmas. Phys. Plasmas **6**(7), 2714–2720 (1999)
40. M. Hnatich, J. Honkonen, M. Jurcisin, Stochastic magnetohydrodynamic turbulence in dimensions $d \geq 2$. Phys. Rev. E **64**, 056411 (2001)
41. R.J. Rivers, *Path Integral Methods in Quantum Field Theory* (Cambridge University Press, Cambridge, 1987)
42. R. Pythian, The functional formalism of classical statistical dynamics. J. Phys. A: Math. Gen. **10**, 777 (1977)
43. D. Hochberg, C. Molina-París, J. Pérez-Mercader, M. Visser, Effective action for stochastic partial differential equations. Phys. Rev. E **60**, 6343 (1999)
44. D. Hochberg, C. Molina-Parıs, J. Perez-Mercader, M. Visser, *Effective Potential for the Massless KPZ Equation*, arXiv/cond-mat/9904391v1 [cond-mat.stat-mech] (1999)
45. D.J. Amit, *Field Theory, the Renormalization Group, and Critical Phenomena* (McGraw-Hill Inc., London, 1978)
46. D. Homeier, G. Münster, *Renormalization Group Analysis of Turbulent Hydrodynamics*, arXiv/1012.0461 (2010)
47. M. Le Bellac, *Quantum and Statistical Field Theory* (Oxford University Press, Oxford, 1991)
48. R.D. Mattuck, *A Guide to Feynman Diagrams in the Many-body Problem*, 2nd edn. (McGraw-Hill International Book Company, New York, 1976)

Chapter 5
Recursion Relations and Fixed Point Analysis

5.1 Hydrodynamics

We begin our discussion of the RG equations by considering neutral fluids with no magnetic interactions. A detailed study can be found in [1, 2]. Our aim is to remark on the results, primarily found in [1] and proceed to a more careful analysis of new terms which are generated under RG.

5.1.1 Model A

5.1.1.1 Differential RG Equations

Consider a randomly forced NS equation with the forcing spectrum:

$$\begin{aligned} D(k) &= D_0 k^2 \qquad |\mathbf{k}| < \Lambda \\ &= 0, \qquad \text{otherwise.} \end{aligned} \tag{5.1.1}$$

By following the notation in [1] we refer to such a model as Model A. By making an appropriate choice of the noise amplitude, chosen in accordance with the fluctuation-dissipation theorem:

$$D_0 = \frac{\nu_0}{\rho} k_B T, \tag{5.1.2}$$

the forcing in the Model A represents a fluid near equilibrium subject to random fluctuations due to the molecular noise. In our prescription of the noise spectrum Model A corresponds to the choice:

$$A_{2n} = 0 \ \ \forall \ \ n \geq 2 \ \text{ and } \ n = 0. \tag{5.1.3}$$

E. Barkhudarov, *Renormalization Group Analysis of Equilibrium and Non-equilibrium Charged Systems*, Springer Theses,
DOI: 10.1007/978-3-319-06154-2_5, © Springer International Publishing Switzerland 2014

The RG equations for this model are:

$$\frac{d\nu}{d\ell} = \left(z - 2 + \tilde{\mathscr{A}}_d \left(\frac{\lambda^2 D_0 \Lambda^\varepsilon}{\nu^3}\right)\right)\nu, \tag{5.1.4}$$

$$\frac{d\lambda}{d\ell} = (\eta - d - 1)\,\lambda, \tag{5.1.5}$$

$$\frac{dD_0}{d\ell} = \left(-d - z + 2\xi - 2 + \tilde{\mathscr{A}}_d \left[\frac{\lambda^2 D_0 \Lambda^\varepsilon}{\nu^3}\right]\right) D_0, \tag{5.1.6}$$

where we have identified $A_2 \equiv D_0$ and

$$\tilde{\mathscr{A}}_d = \mathscr{A}_d \frac{S_d}{(2\pi)^d}, \qquad \mathscr{A}_d = \frac{d^2 - 2}{2d\,(d+2)}. \tag{5.1.7}$$

This set of differential equations can be conveniently re-written using a reduced parameter:

$$x = \lambda D_0^{1/2} \Lambda^{\varepsilon/2} \nu^{-3/2}. \tag{5.1.8}$$

By re-writing differential equations in terms of the flow of the reduced parameter x we can eliminate the exponents $z\,(\ell)$, $\eta(\ell)$ and $\xi\,(\ell)$. This is advantageous because we do not their values at the intermediate stages of the flow, only at fixed points.

The differential equation for x takes the form:

$$\frac{dx}{d\ell} = \frac{1}{2}\,(2 - d)\,x - \tilde{\mathscr{A}}_d x^3. \tag{5.1.9}$$

The behavior of the solution crucially depends on the sign of the linear term. In fact we have already observed the importance of the dimension in determining the phase-space portrait in the study of the d-dimensional Coulomb gas. It is a general feature of all RG calculations. The spatial dimension $d = 2$ is the critical dimension of the system. It separates two regions: when $d > 2$ the solution is driven to zero and when $d < 2$ the system is driven to a stable fixed point:

$$\lim_{\ell \to \infty} x\,(\ell) = 0 \quad \text{for} \quad d \geq 2, \tag{5.1.10}$$

$$\lim_{\ell \to \infty} x\,(\ell) = \left(\frac{\varepsilon}{2\tilde{\mathscr{A}}_{2-\varepsilon}}\right)^{1/2} \quad \text{for} \quad d < 2. \tag{5.1.11}$$

The latter case of $d < 2$ is rather problematic for a description of a fluid. As a result the $d = 1$ Burger's equation is often studied in this region. We will not pursue this study here and focus solely on $d \geq 2$. Although the above finding follow directly from the Eq. (6.1.9) they can also be extracted from the analytic form of the solution:

$$x(\ell) = \frac{x_0 e^{1/2\varepsilon\ell}}{\left(1 + 2\tilde{\mathscr{A}}_d x_0^2 \frac{(e^{\varepsilon\ell}-1)}{\varepsilon}\right)^{1/2}}, \tag{5.1.12}$$

where $\varepsilon = d - 2$. The above solution clearly displays the behavior in Eqs. (5.1.10) and (5.1.11) .[1] The decay of the solution is exponentially fast for $d > 2$. At $d = 2$ the solution falls as:

$$x(\ell) = \frac{x_0}{\left(1 + 2\tilde{\mathscr{A}}_d x_0^2 \ell\right)^{1/2}}. \tag{5.1.13}$$

Once we have established the behavior of the effective coefficient in the limit of $\ell \to \infty$, we are in a position to consider values of the scaling exponents. From the Eq. (5.1.4) it follows that, at a fixed point, the dynamic exponent behaves as:

$$\lim_{\ell\to\infty} z(\ell) = 2 \quad \text{for} \quad d \geq 2, \tag{5.1.14}$$

which is a consequence of the fact that:

$$\frac{d\nu}{d\ell} = 0 \quad \text{at a fixed point.} \tag{5.1.15}$$

Analogous statements can be made about the values of other exponents. At this point it is important to asses the role of other terms which are generated at 1-loop order, which we will explore in the next section.

5.1.1.2 The Role of the Fluctuation-Dissipation Theorem

Model A can be thought of as a description of a fluid near equilibrium. This connection is then supplemented by the fluctuation-dissipation theorem which requires [1]:

$$D_0 = \frac{\nu_0}{\rho} k_B T. \tag{5.1.16}$$

From the differential RG equations, Eqs. (5.1.4)–(5.1.6), we have shown that the reduced coupling $x(\ell) \to 0$ as $\ell \to \infty$, where the definition of this coupling is:

$$x(\ell) = \lambda(\ell) D_0^{1/2}(\ell) \Lambda^{\varepsilon/2} \nu^{-3/2}(\ell). \tag{5.1.17}$$

By requiring the fluctuation-dissipation to hold at any point of the coarse-graining step, i.e. for all ℓ, the reduced coupling should approach its limiting value through

---

[1] The solution in Eq. (5.1.12) differs from [1]. In their chapter the solution is given as $x_0 e^{1/2\varepsilon\ell}\left(1 + 2\tilde{\mathscr{A}}_d x_0^2 \frac{(e^{\varepsilon\ell}-1)}{\varepsilon}\right)^{1/2}$, which differ from our result by a sign of the power $1/2$.

changes of $\lambda(\ell)$, while $D_0(\ell)$ and $\nu(\ell)$ should remain at their initial values throughout the RG trajectory. In order to keep D_0 and ν fixed at their initial values we make the following choice for the dynamic exponent:

$$z(\ell) = 2 - A_d x^2(\ell), \tag{5.1.18}$$

With the above choice for $z(\ell)$ we have $\dot{D}_0 = \dot{\nu} = 0$, given we use the scaling relation:

$$\xi = z + \frac{1}{2}d. \tag{5.1.19}$$

Further, we have to show that under such a choice the perturbation parameter λ does go to zero. The differential RG equation for λ is:

$$\frac{d\lambda(\ell)}{d\ell} = (\eta - d - 1)\lambda(\ell). \tag{5.1.20}$$

With the above choice for z and a restriction that the noise coefficient does not change, we have to conclude that, together with $d + 2z = \xi + \eta$, we can write:

$$\frac{d\lambda(\ell)}{d\ell} = \left(1 - \frac{d}{2} - A_d x^2(\ell)\right)\lambda(\ell). \tag{5.1.21}$$

We can solve this differential equation for λ:

$$\lambda(\ell) = \lambda_0 e^{\ell\left(\frac{2-d}{2}\right) - A_d \int_0^\ell x^2(l)dl}. \tag{5.1.22}$$

Already, we observe that λ goes to zero exponentially fast for $d > 2$. In fact, we can integrate $x^2(\ell)$ exactly to obtain a complete solution:

$$\int_0^\ell dl \frac{x_0^2 e^{-|\varepsilon| l}}{\left(1 + 2x_0^2 \mathscr{A}_d |\varepsilon|^{-1} - 2x_0^2 \mathscr{A}_d |\varepsilon|^{-1} e^{-|\varepsilon| l}\right)}$$
$$= \frac{1}{2\mathscr{A}_d} \ln\left(1 + 2x_0^2 \mathscr{A}_d |\varepsilon|^{-1} - 2x_0^2 \mathscr{A}_d |\varepsilon|^{-1} e^{-|\varepsilon|\ell}\right), \tag{5.1.23}$$

For $\ell \gg 1$ the exponential term can be dropped and for $d \geq 2$ we have:

$$\lambda(\ell) = \begin{cases} \frac{\lambda_0 e^{(2-d/2)\ell}}{\left(1+2x_0^2\mathscr{A}_d|\varepsilon|^{-1}\right)^{1/2}} & \text{for } d > 2, \\ \frac{\lambda_0}{\left(1+2x_0^2\mathscr{A}_d\ell\right)^{1/2}} & \text{for } d = 2. \end{cases}$$

We observe that $\lambda(\ell) \to 0$ as $\ell \to \infty$ in exactly the same way as $x(\ell)$, which is to be expected given above considerations. Note that each RG step does not alter the large-scale behavior of the system. Therefore, we can calculate the desired large scale properties in the form of correlation functions either from the bare theory or from the renormalized theory, the answer would be the same. However, as we have shown, at the macro-scale description the effective parameter λ goes to zero, meaning that the effective theory is described by a linear equation. All terms proportional to λ^n, where $n \geq 1$, become irrelevant [3] and the dynamics is governed by the equation:

$$\partial_t \mathbf{v} = \nu \nabla^2 \mathbf{v} + \mathbf{f}. \tag{5.1.24}$$

5.1.1.3 Homogeneity Relation and Energy Spectrum

An important quantity of interest in fluids is the energy spectrum. Consider the total energy density of the system. It is related to the correlation function via the following equation:

$$E = \frac{1}{2} \mathrm{Tr} \int \frac{d^d \mathbf{k}}{(2\pi)^{d+1}} \frac{d^d \mathbf{q}}{(2\pi)^{d+1}} d\omega d\Omega \left\langle v_i (\mathbf{k}, \omega)\, v_j (\mathbf{q}, \Omega) \right\rangle, \tag{5.1.25}$$

where the trace runs over the indices. This can be conveniently expressed in terms of the correlation functions:

$$\left\langle v_i (\mathbf{k}, \omega)\, v_j (\mathbf{q}, \Omega) \right\rangle = G_{ij} (\mathbf{k}, \omega)\, \delta (\mathbf{k} + \mathbf{q})\, \delta (\omega + \Omega), \tag{5.1.26}$$

which gives:

$$E = \frac{1}{2} \mathrm{Tr} \int \frac{d^d \mathbf{k}}{(2\pi)^{d+1}} d\omega G_{ij} (\mathbf{k}, \omega). \tag{5.1.27}$$

We can now define the energy spectrum density as:

$$E (k) = \frac{1}{2} \frac{S_d}{(2\pi)^{d+1}} k^{d-1} \mathrm{Tr} \int d\omega G_{ij} (\mathbf{k}, \omega). \tag{5.1.28}$$

We are interested in how does this energy spectrum density scale with k. To investigate that we consider how the correlation function behave under fast modes elimination. The RG transformation does not alter the behavior of the long wave modes. Thus, we can calculate the correlation function either from the bare theory or from the renormalized theory with renormalized coefficients. Mathematically, this statement is equivalent to saying that the correlation function is homogeneous:

$$G_{ij} (\mathbf{k}, \omega, \lambda_0) = b^{-\chi(\ell)} G_{ij} \left(e^{\ell} \mathbf{k}, e^{\alpha(\ell)} \omega, \bar{\lambda} (\ell) \right). \tag{5.1.29}$$

We can derive the scaling rules for the correlation function by considering the following:

$$\begin{aligned}\int d\mathbf{q}d\Omega \langle v_i(\mathbf{k},\omega)\, v_j(\mathbf{q},\Omega)\rangle &= b^{-z-d}\int d\mathbf{q}'d\Omega' \langle v_i\left(b^{-1}\mathbf{k}', b^{-z}\omega'\right) v_j\left(b^{-1}\mathbf{q}', b^{-z}\Omega'\right)\rangle \\ &= b^{2\eta-z-d}\int d\mathbf{q}'d\Omega' \langle v_i\left(\mathbf{k}',\omega'\right) v_j\left(\mathbf{q}',\Omega'\right)\rangle \\ &= e^{2\eta-z-d} G_{ij}\left(e^{\ell}\mathbf{k}, e^{\alpha(\ell)}\omega, \bar{\lambda}(\ell)\right). \end{aligned} \tag{5.1.30}$$

The last line follows from the rescaling of the fields and re-expressing the correlation function in terms of rescaled variables. We can exploit some of the scaling relations used in the previous section to arrive to the following equation:

$$G_{ij}(\mathbf{k},\omega,\lambda_0) = e^{\int_0^{\ell} z(\ell')d\ell'} G_{ij}\left(e^{\ell}\mathbf{k}, e^{\alpha(\ell)}\omega, \bar{\lambda}(\ell)\right). \tag{5.1.31}$$

The correlation functions scales like the frequency exponent, therefore up to the leading order in k we have that the energy spectrum goes as:

$$E(k) \propto k^{d-1}. \tag{5.1.32}$$

This result holds for $d \geq 2$.

5.1.2 Galilean Invariance

So far we have concluded that the asymptotic form of the equation of motion is a linear equation:

$$\partial_t \mathbf{v} = \nu \nabla^2 \mathbf{v} + \mathbf{f}, \tag{5.1.33}$$

which is a consequence of a fluctuation-dissipation theorem. However, the above equation does not preserve the symmetry of the original Navier-Stokes equation, the Galilean symmetry. It will be shown, based on the work of [4, 5], that, in general, the Galilean symmetry does not put any constraints on the vertex renormalization (the non-linear term) because, as in the Model A case, one often is interested in velocity fluctuations around the mean value of the velocity field. The velocity fluctuations are in fact velocity differences, so they are invariant under the Galilean transformation from the outset.

5.1.2.1 The Symmetry of the Equation of Motion

Consider the Navier-Stokes equation with the pressure term explicitly present:

$$\frac{\partial V_i}{\partial t} + V_j \frac{\partial V_i}{\partial x_j} = -\frac{\partial P}{\partial x_i} + \nu_0 \nabla^2 V_i, \quad \nabla \cdot \mathbf{V} = 0. \tag{5.1.34}$$

Taking divergence of the above equation leads to:

$$\nabla^2 P = -\frac{\partial^2}{\partial x_i \partial x_j} \left(V_j V_i \right). \tag{5.1.35}$$

The Galilean transformation is a transformation of variables of the form:

$$\mathbf{x}' = \mathbf{x}'\left(\mathbf{x}, t\right), \quad t' = t'\left(\mathbf{x}, t\right), \tag{5.1.36}$$

such that

$$\mathbf{x}' = \mathbf{x} - \lambda \mathbf{c} t, \quad t' = t. \tag{5.1.37}$$

The derivatives in the equation of motion then transform:

$$\frac{\partial}{\partial x_i} = \delta_{ij} \frac{\partial}{\partial x'_j}, \quad \frac{\partial}{\partial t} = -\lambda c_i \frac{\partial}{\partial x'_i} + \frac{\partial}{\partial t'}. \tag{5.1.38}$$

In order to preserve the equation of motion under such transformation we require that the field transforms as

$$V'_i\left(\mathbf{x}', t'\right) = V_i\left(\mathbf{x}' + \lambda \mathbf{c} t', t'\right) - \lambda c_i. \tag{5.1.39}$$

By explicitly performing the transformation we arrive at the result:

$$\begin{aligned} -\lambda c_j \frac{\partial V'_i\left(\mathbf{x}', t'\right)}{\partial x'_j} + \frac{\partial V'_i\left(\mathbf{x}', t'\right)}{\partial t'} + V'_j\left(\mathbf{x}', t'\right) \frac{\partial V'_i\left(\mathbf{x}', t'\right)}{\partial x'_j} + \lambda c_j \frac{\partial V'_i\left(\mathbf{x}', t'\right)}{\partial x'_j} \\ = -\frac{\partial P'}{\partial x'_i} + \nu_0 \nabla^2 V'_i\left(\mathbf{x}', t'\right). \end{aligned} \tag{5.1.40}$$

Now, by setting $\lambda = 1$ we obtain the transformed equation in a new reference frame:

$$\frac{\partial V'_i\left(\mathbf{x}', t'\right)}{\partial t'} + V'_j\left(\mathbf{x}', t'\right) \frac{\partial V'_i\left(\mathbf{x}', t'\right)}{\partial x'_j} = -\frac{\partial P'}{\partial x'_i} + \nu_0 \nabla^2 \left(V'_i\left(\mathbf{x}', t'\right)\right). \tag{5.1.41}$$

The equation is invariant under a Galilean transformation. From the above considerations we conclude that if under the RG transformation the coefficient of the non-linear term changes then the Galilean invariance will be broken.

5.1.2.2 Reynolds Decomposition

The linear equation which results from a repeated RG transformation appears to violate Galilean invariance. However, recall that Model A is a theory formulated to describe equilibrium fluctuations of the fluid. Thus, the equation derived expresses the fluctuations of the velocity field around some mean value. This point was heavily emphasized by [4]. Here, we would like to explicitly illustrate this idea.

Consider expressing the velocity field $\mathbf{V}$ in terms of its mean value and fluctuations around the mean:

$$\mathbf{V} = \mathbf{U} + \mathbf{u}, \tag{5.1.42}$$

such that

$$\langle \mathbf{V} \rangle = \mathbf{U}. \tag{5.1.43}$$

The $\langle \cdot \rangle$ is an ensemble average. Analogously we decompose the pressure:

$$P = P_{av} + p. \tag{5.1.44}$$

Now, consider field transformation under Galilean transformation while taking into account Reynolds decomposition:

$$U'_i\left(\mathbf{x}', t'\right) + u'_i\left(\mathbf{x}', t'\right) = U_i\left(\mathbf{x}' + \lambda \mathbf{c}t', t'\right) + u_i\left(\mathbf{x}' + \lambda \mathbf{c}t', t'\right) - c_i. \tag{5.1.45}$$

We can derive how each part of the decomposed velocity field transforms by taking the ensemble average of the above expression. We conclude:

$$U'_i\left(\mathbf{x}', t'\right) = U_i\left(\mathbf{x}' + \lambda \mathbf{c}t', t'\right) - c_i, \tag{5.1.46}$$

which implies:

$$u'_i\left(\mathbf{x}', t'\right) = u_i\left(\mathbf{x}' + \lambda \mathbf{c}t', t'\right). \tag{5.1.47}$$

The above is a field transformation for the fluctuating part of the velocity field. If we assume that the mean part is a constant in space and time we can write the equation of motion for the fluctuations:

$$\frac{\partial u_i}{\partial t} + U_j \frac{\partial u_i}{\partial x_j} + \frac{\partial\left(u_j u_i\right)}{\partial x_j} = -\frac{\partial p}{\partial x_i} + \nu_0 \nabla^2 u_i. \tag{5.1.48}$$

This equation is Galilean invariant under:

$$u'_i\left(\mathbf{x}', t'\right) = u_i\left(\mathbf{x}' + \lambda \mathbf{c}t', t'\right), \quad U'_i = U_i - c_i, \quad \frac{\partial}{\partial t} = -c_i \frac{\partial}{\partial x'_i} + \frac{\partial}{\partial t'}. \tag{5.1.49}$$

Further note that if we move into the co-moving frame of reference the second term in Eq. (6.1.49) can be eliminated to give:

$$\frac{\partial u_i}{\partial t} + \frac{\partial\left(u_j u_i\right)}{\partial x_j} = -\frac{\partial p}{\partial x_i} + \nu_0 \nabla^2 u_i. \tag{5.1.50}$$

It must be noted that now, the equations are already Galilean invariant and there is no way of directly testing it without moving away from the co-moving frame. The key is to note that Eq. (5.1.34), our original equation of motion, bears the same form. However they represent two different fields and therefore they do not transform in the same way.

5.1.2.3 Vertex Renormalization

Consider the Galilean transformations in Fourier space, where we define the Fourier transform as:

$$f(\mathbf{k}, t) = \int d\mathbf{x} f(\mathbf{x}, t) e^{-i\mathbf{k}\cdot\mathbf{x}}. \tag{5.1.51}$$

The fields are transformed as

$$\tilde{V}_i'(\mathbf{k}, t') = e^{i\lambda\mathbf{k}\cdot\mathbf{c}t'} \tilde{V}_i(\mathbf{k}, t') - c_i\delta(\mathbf{k}), \tag{5.1.52}$$

while the pressure term transform as:

$$P(\mathbf{x}' + \lambda\mathbf{c}t', t') = P'(\mathbf{x}', t'). \tag{5.1.53}$$

Hence, in Fourier space we have:

$$\tilde{P}'(\mathbf{k}, t') = e^{i\lambda\mathbf{k}\cdot\mathbf{c}t} \tilde{P}(\mathbf{k}, t'). \tag{5.1.54}$$

The Navier-Stokes equation in Fourier space takes the form from:

$$\begin{aligned}&\frac{\partial \tilde{V}_i(\mathbf{k}, t)}{\partial t} + ik_j \int \frac{d\mathbf{q}}{(2\pi)^d} \tilde{V}_j(\mathbf{q}, t) \tilde{V}_i(\mathbf{k} - \mathbf{q}, t)\\ &= -ik_i \tilde{P}(\mathbf{k}, t) - k^2 \nu_0 \tilde{V}_i(\mathbf{k}, t).\end{aligned} \tag{5.1.55}$$

Consider how fields transform under Reynolds decomposition in wave number space:

$$\tilde{V}_i(\mathbf{k}, t') = \tilde{U}_i(\mathbf{k}, t') + \tilde{u}_i(\mathbf{k}, t'). \tag{5.1.56}$$

It follows that the transformation rules are:

$$\tilde{U}_i'(\mathbf{k}, t') = e^{i\lambda\mathbf{k}\cdot\mathbf{c}t'} U_i(\mathbf{k}, t') - c_i\delta(\mathbf{k}), \tag{5.1.57}$$

and

$$\tilde{u}_i'(\mathbf{k}, t') = e^{i\lambda\mathbf{k}\cdot\mathbf{c}t'} \tilde{u}_i(\mathbf{k}, t'). \tag{5.1.58}$$

The equation of motion for the fluctuating fields can be readily obtained (assuming constant average velocity independent of space-time coordinate $U_i(\mathbf{k}, t) = K_i \delta(\mathbf{k})$):

$$\frac{\partial \tilde{u}_i(\mathbf{k}, t)}{\partial t} + ik_j K_j \tilde{u}_i(\mathbf{k}, t) + ik_j \int \frac{d\mathbf{q}}{(2\pi)^d} \tilde{u}_j(\mathbf{q}, t) \tilde{u}_i(\mathbf{k} - \mathbf{q}, t)$$
$$= -ik_i \tilde{p}(\mathbf{k}, t) - k^2 \nu_0 \tilde{u}_i(\mathbf{k}, t) . \tag{5.1.59}$$

At last we can concentrate on the consequences of the Galilean invariance on RG in the laboratory frame. Let us split the fluctuating field into fast and slow Fourier components:

$$\tilde{u}_i(\mathbf{k}, t) = \tilde{u}_i^>(\mathbf{k}, t) + \tilde{u}_i^<(\mathbf{k}, t) . \tag{5.1.60}$$

Substitute this into the equation of motion:

$$\begin{aligned}
-ik_i \tilde{p}^<(\mathbf{k}, t) - k^2 \nu_0 \tilde{u}_i^<(\mathbf{k}, t) = {} & \frac{\partial \tilde{u}_i^<(\mathbf{k}, t)}{\partial t} + ik_j K_j \tilde{u}_i^<(\mathbf{k}, t) \\
& + ik_j \theta\left(k - \Lambda b^{-1}\right) \left[\int \frac{d\mathbf{q}}{(2\pi)^d} \tilde{u}_j^<(\mathbf{q}, t) \tilde{u}_i^<(\mathbf{k} - \mathbf{q}, t) \right. \\
& + ik_j \int \frac{d\mathbf{q}}{(2\pi)^d} \tilde{u}_j^>(\mathbf{q}, t) \tilde{u}_i^<(\mathbf{k} - \mathbf{q}, t) \\
& + ik_j \int \frac{d\mathbf{q}}{(2\pi)^d} \tilde{u}_j^<(\mathbf{q}, t) \tilde{u}_i^>(\mathbf{k} - \mathbf{q}, t) \\
& \left. + ik_j \int \frac{d\mathbf{q}}{(2\pi)^d} \tilde{u}_j^>(\mathbf{q}, t) \tilde{u}_i^>(\mathbf{k} - \mathbf{q}, t) \right] .
\end{aligned} \tag{5.1.61}$$

We assume that we have come up with some procedure which would give us a coarse grained description of slow modes. This would generally include corrections to the viscosity and the vertex:

$$\frac{\partial \tilde{u}_i^<(\mathbf{k}, t)}{\partial t} + ik_j K_j \tilde{u}_i^<(\mathbf{k}, t) + ik_j \theta\left(k - \Lambda b^{-1}\right) \lambda(\mathbf{k}) \int \frac{d\mathbf{q}}{(2\pi)^d} \tilde{u}_j^<(\mathbf{q}, t) \tilde{u}_i^<(\mathbf{k} - \mathbf{q}, t)$$
$$= -ik_i \tilde{p}^<(\mathbf{k}, t) - k^2 \nu_0(\mathbf{k}) \tilde{u}_i^<(\mathbf{k}, t) . \tag{5.1.62}$$

The way our equation transforms puts no restriction on the vertex renormalization as they are still Galilean invariant. The additional term introduced from the time derivative is canceled by transformation of K_j.

The above considerations illustrate that the linear theory which governs macroscopic fluctuations in an equilibrium fluid is consistent with the original symmetries of the problem. Further, one should not invoke Galilean symmetry to justify non-renormalizability of the vertex function. In fact, the above considerations can be made more rigorous using the functional integral formulation by studying consequences of continuous symmetries.

5.1.2.4 Ward-Takahashi Identities and Field-Theoretic Approach

In the field theory formalism continuous symmetries of the system provide a natural mechanism to study relationships between verticies [6, 7]. The resultant equations are valid to all orders in perturbation theory. Here, we would like to touch upon such a development in the context of the Navier-Stokes equation [5, 10].

By definition, a symmetry operation implies that the action remains invariant under the transformation. A continuous symmetry can be made infinitesimally small, so let us consider what this means for the Galilean transformations. The generating functional can be readily derived from the full MHD generating functional. It has the form:

$$\mathscr{Z}[\mathbf{J},\Sigma]=\int\mathscr{D}[\mathbf{u}]\mathscr{D}[\sigma]\exp\left(-S[\mathbf{u},\sigma]+\int d\mathbf{x}dt\mathbf{J}\cdot\mathbf{u}\right.$$
$$\left.+\int d\mathbf{x}dt\mathbf{J}\cdot\mathbf{K}+\int d\mathbf{x}dt\Sigma\cdot\sigma\right), \tag{5.1.63}$$

where the action is defined as:

$$S[\mathbf{u},\sigma]=\int d\mathbf{x}\int dt\left[\frac{1}{2}\int d\mathbf{y}\sigma_i(\mathbf{x},t)D_{ij}(\mathbf{x}-\mathbf{y})\sigma_j(\mathbf{y},t)\right.$$
$$\left.-i\sigma_k\left(\frac{\partial u_k}{\partial t}+K_j\frac{\partial u_k}{\partial x_j}+\lambda_0P_{kj}(\nabla)\frac{\partial(u_iu_j)}{\partial x_i}-\nu_0\nabla^2u_k\right)\right]. \tag{5.1.64}$$

K_j is just a constant which represents the average fields and u_j are fluctuations about it. Under a symmetry transformation the action remains invariant by definition. The only change comes from the products of physical fields with the source fields. Consider infinitesimal transformations of the field under the action of the Galilean transformation:

$$u_i(\mathbf{x}+\delta\mathbf{c}t,t)=u_i(\mathbf{x},t)+\delta\mathbf{c}_jt\nabla_ju_i(\mathbf{x},t), \tag{5.1.65}$$

$$\sigma_i(\mathbf{x}+\delta\mathbf{c}t,t)=\sigma_i(\mathbf{x},t)+\delta\mathbf{c}_jt\nabla_j\sigma_i(\mathbf{x},t). \tag{5.1.66}$$

This would generate an infinitesimal change in the generating functional:

$$\delta\mathscr{Z}[\mathbf{J},\Sigma]=\left\langle\int d\mathbf{x}dt\left[J_i\left(\delta\mathbf{c}_jt\nabla_ju_i(\mathbf{x},t)\right)-\delta c_iJ_i+\Sigma_i\delta\mathbf{c}_jt\nabla_j\sigma_i(\mathbf{x},t)\right]\right\rangle=0, \tag{5.1.67}$$

which can be re-written as

$$\int d\mathbf{x}dt\left[tJ_i\nabla_j\frac{\delta}{\delta J_i(\mathbf{x},t)}-J_j+t\Sigma_i\nabla_j\frac{\delta}{\delta\Sigma_i(\mathbf{x},t)}\right]\mathscr{Z}[\mathbf{J},\Sigma]=0, \tag{5.1.68}$$

by exploiting the fact that $\mathscr{Z}[\mathbf{J}, \Sigma]$ is a generating functional and therefore all the fields can be expressed as derivatives with respect to the corresponding source field. An important detail is that the mean field $\mathbf{K}$ is treated on equal grounds with $\mathbf{u}$ and hence they have the same conjugate field. So if we try to compute the average field $\langle \mathbf{V} \rangle$ we would obtain $\mathbf{K}$, which is consistent with the Reynolds decomposition.

To this end we have derived a functional equation which provides us with an infinite number of relations relating various correlation functions. These are generated by taking various derivatives of Eq. (5.1.68). The set of equations generated as a result are referred to as Ward-Takahashi identities. However, as it stands, these equations are cumbersome since, by working with the generating functional $\mathscr{Z}[\mathbf{J}, \Sigma]$, we would generate all kinds of Feynman graphs: connected, disconnected, 1-particle-reducible and 1-particle-irreducible. Instead of working with $\mathscr{Z}[\mathbf{J}, \Sigma]$, one can work with a "free energy" analogue. Consider the quantity:

$$\mathscr{F}[\mathbf{J}, \Sigma] = \ln \mathscr{Z}[\mathbf{J}, \Sigma]. \tag{5.1.69}$$

The new functional satisfies:

$$\int d\mathbf{x}dt \left[tJ_i \nabla_j \frac{\delta \mathscr{F}[\mathbf{J}, \Sigma]}{\delta J_i(\mathbf{x}, t)} - J_j + t\Sigma_i \nabla_j \frac{\delta \mathscr{F}[\mathbf{J}, \Sigma]}{\delta \Sigma_i(\mathbf{x}, t)} \right] = 0. \tag{5.1.70}$$

Even though we have reduced the number of graphs we have to consider, this is still not the most convenient form to use because both 1-particle-reducible and 1-particle-irreducible diagrams are generated. Instead one chooses to work with the Legendre transform of $\mathscr{F}[\mathbf{J}, \Sigma]$ [8, 9]. It allows us to work exclusively in terms of 1-particle-irreducible diagrams. Further, this form is the most useful to us since under RG the leading corrections to the vertex renormalization λ and to the viscosity ν are represented by 1-particle-irreducible diagrams. The connection between the two is most conveniently expressed through the Legendre transform of $\mathscr{F}$, which is also known as the vertex function.

To introduce the Legendre transform we begin by defining new field variables:

$$\frac{\delta \mathscr{F}[\mathbf{J}, \Sigma]}{\delta J_i} = \overline{\mathbf{u}} + \mathbf{K}, \quad \frac{\delta \mathscr{F}[\mathbf{J}, \Sigma]}{\delta \Sigma_i} = \overline{\sigma}_i. \tag{5.1.71}$$

Then we define a functional:

$$\Gamma[\overline{\mathbf{u}} + \mathbf{K}, \overline{\sigma}] = -\mathscr{F}[\mathbf{J}, \Sigma] + \int d\mathbf{x} \int dt\, (\mathbf{J} \cdot (\overline{\mathbf{u}} + \mathbf{K}) + \Sigma \cdot \overline{\sigma}), \tag{5.1.72}$$

$$\frac{\delta \Gamma[\overline{\mathbf{u}} + \mathbf{K}, \overline{\sigma}]}{\delta (\overline{u} + K)_i} = J_i, \quad \frac{\delta \Gamma[\overline{\mathbf{u}} + \mathbf{K}, \overline{\sigma}]}{\delta \overline{\sigma}_i} = \Sigma_i. \tag{5.1.73}$$

These sets of definitions implies that Γ satisfies the equation:

$$\int d\mathbf{x}dt\left[t\frac{\delta\Gamma}{\delta\left(\overline{u}+K\right)_i}\nabla_j\left(\overline{u}_i+K_i\right)+t\frac{\delta\Gamma}{\delta\overline{\sigma}_i}\nabla_j\overline{\sigma}_i-\frac{\delta\Gamma}{\delta\left(\overline{u}+K\right)_j}\right]=0. \quad (5.1.74)$$

Consider differentiation the above expression with respect to $(\overline{u}+K)_k\left(y,t'\right)$. This results in an expression:

$$\begin{aligned}
0=&\int d\mathbf{x}dt\frac{\delta^2\Gamma}{\delta\left(\overline{u}+K\right)_k\left(y,t'\right)\delta\left(\overline{u}+K\right)_i\left(\mathbf{x},t\right)}\nabla_j\left(\overline{u}_i+K_i\right)\\
&+\int d\mathbf{x}dt\frac{\delta\Gamma}{\delta\left(\overline{u}+K\right)_i\left(\mathbf{x},t\right)}\nabla_j\delta_{ki}\delta\left(\mathbf{x}-\mathbf{y}\right)\delta\left(t-t'\right)\\
&+\int d\mathbf{x}dt\frac{\delta^2\Gamma}{\delta\overline{\sigma}_i\left(\mathbf{x},t\right)\delta\left(\overline{u}+K\right)_k\left(\mathbf{y},t'\right)}\nabla_j\overline{\sigma}_i\\
&-\int d\mathbf{x}dt\frac{\delta^2\Gamma}{\delta\left(\overline{u}+K\right)_k\left(\mathbf{y},t'\right)\delta\left(\overline{u}+K\right)_j\left(\mathbf{x},t\right)}. \quad (5.1.75)
\end{aligned}$$

Further, let us differentiate the last expression with respect to $\frac{\delta}{\delta\sigma_s(z,t'')}$. As with other generating functionals we set the fields $\overline{\sigma}$ and $(\overline{u}+K)$ to zero at the end of the desired calculation. Note setting $(\overline{u}+K)=0$ to zero corresponds to the isotropic homogeneous flow i.e. no symmetry breaking, while $(\overline{u}+K)=K$, corresponds to a broken symmetry case, since the flow has a preferred direction K. As we have established in the previous section, in order to be able to test for Galilean invariance one has to be out of the co-moving frame of reference, the laboratory frame. Hence we choose to consider the relationships in a broken symmetry case. The final relationship reads:

$$\begin{aligned}
0=&-\delta c_i\int d\mathbf{x}dt\frac{\delta^3\Gamma}{\delta\sigma_s\left(\mathbf{z},t''\right)\delta\left(\overline{u}+K\right)_k\left(\mathbf{y},t'\right)\delta\left(\overline{u}_i+K_i\right)\left(\mathbf{x},t\right)}\\
&-t'\delta\mathbf{c}\cdot\nabla_{\mathbf{y}}\frac{\delta^2\Gamma}{\delta\sigma_s\left(\mathbf{z},t''\right)\delta\left(\overline{u}+K\right)_k\left(\mathbf{y},t'\right)}\\
&-t''\delta\mathbf{c}\cdot\nabla_{\mathbf{z}}\frac{\delta^2\Gamma}{\delta\left(\overline{u}+K\right)_k\left(\mathbf{y},t'\right)\delta\overline{\sigma}_s\left(\mathbf{z},t''\right)}. \quad (5.1.76)
\end{aligned}$$

In effect we have expanded the vertex function around the mean flow. We can further simplify the above expression to have an exact correspondence with [5] by noting that the vertex function is translationally invariant as a consequence of the assumption that we are in a homogeneous flow. Consider a more compact notation:

$$\frac{\delta^2\Gamma}{\delta\sigma_n\left(\mathbf{z},t''\right)\delta\left(\overline{u}_s+K_s\right)\left(\mathbf{y},t'\right)}=\Gamma_{ns}^{[1,1]}\left(\mathbf{z},t'';\mathbf{y},t'\right), \quad (5.1.77)$$

then it follows from translational invariance that:

$$\Gamma_{ns}^{[1,1]}\left(\mathbf{z}, t''; \mathbf{y}, t'\right) = \Gamma_{ns}^{[1,1]}\left(\mathbf{z}+\mathbf{a}, t''+T; \mathbf{y}+\mathbf{a}, t'+T\right). \tag{5.1.78}$$

By choosing $\mathbf{a} = -\mathbf{z}$ and $T = -t''$ we observe:

$$\Gamma_{ns}^{[1,1]}\left(\mathbf{z}, t''; \mathbf{y}, t'\right) = \Gamma_{ns}^{[1,1]}\left(\mathbf{0}, 0; \mathbf{y}-\mathbf{z}, t'-t''\right). \tag{5.1.79}$$

Hence, the vertex function is a function of temporal and spatial difference only. It follows:

$$\delta c_i \int d\mathbf{x} dt \Gamma_{ski}^{[1,2]}\left(\mathbf{y}-\mathbf{z}, t'-t''; \mathbf{x}-\mathbf{z}, t-t''\right) = \left(t''-t'\right)\delta\mathbf{c}\cdot\nabla_{\mathbf{y}}\Gamma_{sk}^{[1,1]}\left(\mathbf{y}-\mathbf{z}, t'-t''\right). \tag{5.1.80}$$

The above result is valid to all order in the perturbation theory. In other words, it is exact. Further, we can clearly exploit the connection between the two-point and a three-point vertex from the above example which is what was done in [5]. In practice one often chooses to work in Fourier space since the problem of diverging integrals is more apparent through Fourier modes. Thus, in Fourier space,one derives:

$$-k_i \frac{\partial}{\partial\omega}\Gamma_{ks}^{[1,1]}(-\mathbf{k}, -\omega) = \Gamma_{ski}^{[1,2]}(\mathbf{k}, \omega; \mathbf{q}, \Omega). \tag{5.1.81}$$

We stress that these results are exact, so one can exploit them at any order of the perturbative expansion. As a consequence of the identity Eq. (5.1.82), it can be shown that vertex corrections are only partially constrained. Specifically, only in the case of zero momentum transfer, as a consequence of the Galilean symmetry, the vertex does not renormalize.

An important aspect of the above derivation of the Ward identities is the fact that the action is formulated explicitly in the laboratory frame of reference, whereby the mean flow velocity is explicitly present. As was pointed out by McComb [4], in order to establish Galilean invariance one has to be out of the co-moving frame. This is exactly the reason for formulating the functional integral through Reynolds decomposition.

5.1.3 Relevant, Marginal and Irrelevant Variables

5.1.3.1 Higher Order Corrections in External Momentum

The first aspect which we would like draw attention to are the loop corrections themselves. We have performed an expansion in external momentum and kept only the lowest order terms. To asses the role of higher order terms let us go back to a graphical representation of our results. Up to 1-loop order the corrections to the noise covariance can be expressed diagramatically:

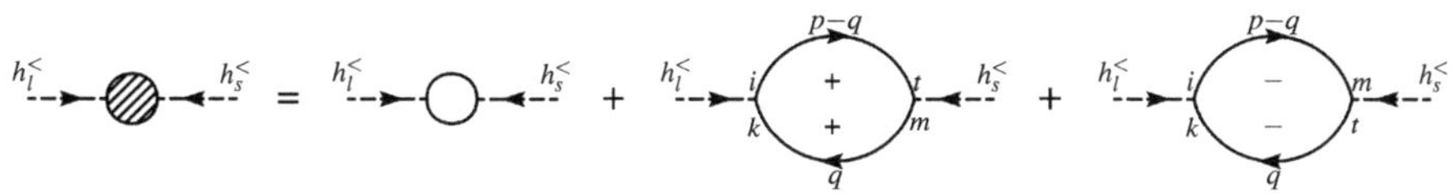

Fig. 5.1 Corrections to noise covariance up to 1-loop order

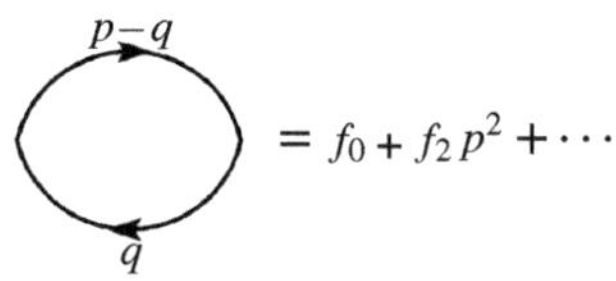

Fig. 5.2 Taylor expansion of the loop in the external momentum

Figure 5.1 refers to Elssaser variables. In the limit of the Navier-Stokes equation, the loop correction is identical in the last two terms, which accounts for the symmetry factor of 2. Consider the Taylor expansion of the loop in the external momentum:

Every term in this Taylor expansion is proportional to:

$$f_{2i} \propto D_0 x_0 \int_{\Lambda b^{-1}}^{\Lambda} k^{d-3} dk \quad \forall \quad i \in \{0, 1, \dots\} \tag{5.1.82}$$

The above proportionality relation identifies regimes where the perturbation theory is applicable. This crucially depends on the dimension. For $d > 2$ the integral is convergent in the limit of $b \to \infty$. For $d = 2$ perturbation theory diverges logarithmically. Consider what happens upon rescaling of the cut-off and the corresponding rescaling of the fields and other variables. The left-hand side of the diagrammatic equation in Fig. 5.2 can be written in terms of its scaling factors as:

$$D^b(k) = b^{-d-z+2\xi-2} k^2 \left(D_0 + f_0(D_0, x_0) + b^{-2} k^2 f_2(D_0, x_0) + \cdots \right), \tag{5.1.83}$$

where D_0 is a 0-loop order coefficient of the noise covariance. D_b denotes a general form of the noise covariance after a shell of thickness b of fast Fourier modes has been eliminated. By explicitly looking at the loop integrals, one observes that it depends only on the effective coupling and D_0. Consecutive mode elimination leads to:

$$D^{b(n+1)}(k) = b^{(n+1)(-d-z+2\xi-2)} k^2 D_0^n \left(1 + f_0(x_n) + b^{-2} k^2 f_2(x_n) + \cdots \right) \tag{5.1.84}$$

By looking at a fixed point solution to lowest order in the expansion, we can asses the role of higher order terms. A fixed point would correspond to the following equation:

$$D_0^* = b^{(n+1)(-d-z+2\xi-2)} D_0^* (1 + f_0(x_n)) \quad \forall \quad n. \tag{5.1.85}$$

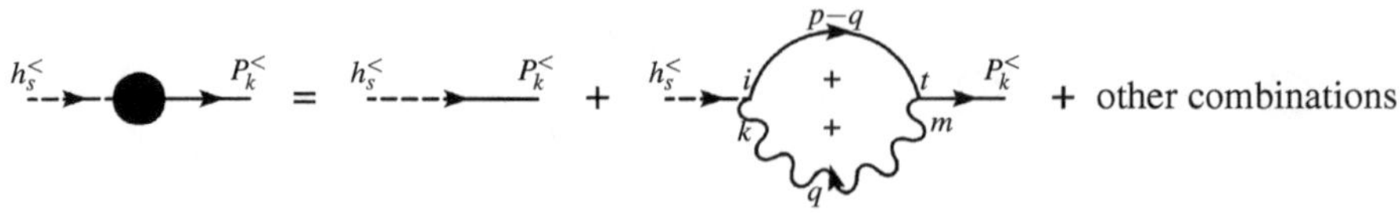

Fig. 5.3 Corrections to γ_+ coefficient in Elssaser variables up to 1-loop order

We have shown that $x(\ell) \to 0$ which in turn suggests that $f_0(x_n) \to 0$ as $\ell \to \infty$. Thus, at a fixed point, the following relationship must hold:

$$2\xi = 2 + d + z. \tag{5.1.86}$$

As a consequence, higher-order corrections are irrelevant. In particular, terms of the form p^4 and higher powers of $\mathbf{p}$ in the noise covariance are irrelevant. Also note that even at $d = 2$ although f_0 goes as $\ln b$ as $b \to \infty$, the effective coupling goes as $1/\ell$ which cancels the integral. In that sense higher-order terms decay even faster (Fig. 5.3).

Similar considerations can be applied to the response function:

By considering the Taylor expansion of the loop integral, after rescaling we can write:

$$\gamma_+^b(k) = b^{z-2}\gamma_+^0\left(k^2 + g_0(x_0)k^2 + b^{-2}g_2(x_0)k^4 +\right), \tag{5.1.87}$$

where $g_0 + g_2k^2 + \cdots$ is a Taylor expansion in external momenta of all combined loop diagrams. It is apparent that on the basis of the discussion of the noise covariance, higher-order terms are irrelevant near a fixed point. The same is true of higher-order corrections to λ. All of the above considerations are simplified by the fact that we have a single effective coupling on which the behavior of a loop diagram depends. Also, in both cases, for $\gamma_\pm$ and λ, the loop diagram is once again proportional to $x_0 \int_{\Lambda b^{-1}}^{\Lambda} k^{d-3}dk$ to all orders in the external momentum. So everything remains finite when we take the limit $b \to \infty$.

5.1.3.2 Connected Graphs

Our formalism within the RG calculation allowed us to ignore terms which are disconnected because they all cancel. Within the diagrammatic expansion technique there exists one further classification of diagrams into 1-particle-reducible and 1-particle-irreducuble diagrams. For example, consider a diagram:

We can cut the middle line to break the diagram into two irreducible diagrams. The effect of such terms has to be assessed within our RG framework.

The graphical representation in Fig. 5.4 represents a multiplicative noise term. There are no loops present, so we do not have to carry out any integration over fast Fourier modes. We are only concerned with restoring the momentum cut-off to Λ.

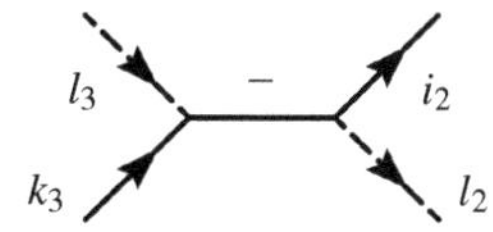

Fig. 5.4 An example of a reducible diagram, which is generated at 1-loop order

Fig. 5.5 A term which is generated under RG at 1-loop order

Under such rescaling transformation the graph scales as b^{-d+2}, which means that in the regime $d > 2$ this term is irrelevant as $\ell \to \infty$. However, at $d = 2$ this term is marginal. We cannot asses its magnitude and without higher order analysis we cannot establish whether $x\,(\ell \to \infty) = 0$ is a true fixed point.

Consider a reducible diagram which is generated under the RG in Fig. 5.5. Under the scale transformation this graph scales as b^0 for all values of d. As has been mentioned, the effect of marginal terms cannot be assessed without analyzing higher-order terms. This strategy is fine as long as there is a finite number of marginal terms in the expansion. However as has been shown by [3] there are infinitely many such terms even at the level of 1-loop.

To extend the scaling argument for the marginal terms to all order in λ, consider how an arbitrarily large graph of order λ^n behaves under scaling [3]. In general we can write how various components of the diagram depicted in Fig. 5.6 scale:

$$\begin{aligned}
\text{Scaling of integrals} &= -2n\,(d+z),\\
\text{Derivatives} &= -n,\\
\text{Correlation function} &= (n-1)\,(2+d+z),\\
\text{Physical fields} &= (n+1)\eta,\\
\text{Noise} &= \xi.
\end{aligned}$$

By adding all the scaling contributions one can convince himself that they add up to zero, which means that there are infinitely many marginal terms generated in our expansion. This has a destructive effect on the whole RG procedure, since a simple higher order analysis is impossible and therefore we cannot truly establish the existence of fixed points in our analysis and would require us to consider an infinite parameter space to specify the action. In the context of the randomly-forced Navier-Stokes equation, the existence of infinitely many marginal terms, under a particular choice of exponents, was first pointed out by [3]. Inspite of this apparent difficulty, the scaling which is motivated through the fluctuation-dissipation theorem, such terms are irrelevant.

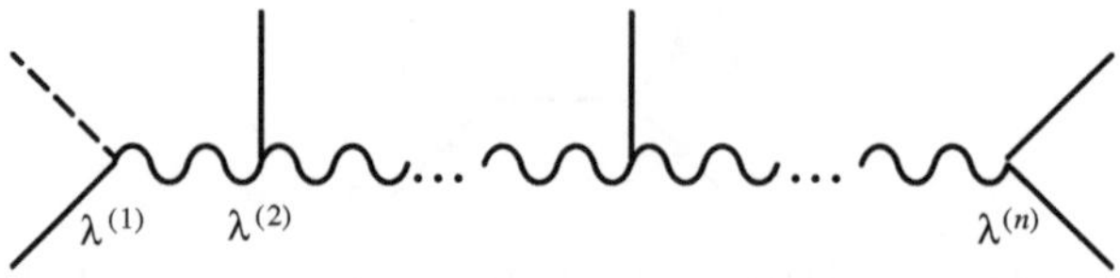

Fig. 5.6 A marginal term of order λ^n which is generated under RG at 1-loop order

5.1.3.3 Terms of Higher Order in λ

It is worth reminding the reader that our calculation is done within a loop approximation. Thus, we should consider all terms which are generated at 1-loop order. This requires us in principle to go to arbitrary large order in λ. We have already seen such an example of 0-loop term, Fig. 5.6. For the Model A we have identified the effective expansion coefficient $x(\ell)$, which goes to zero for $d \geq 2$. It therefore follows that correction of higher order in λ^n are negligible in the limit of $\ell \to \infty$ for the Model A analysis.

5.2 Magnetohydrodynamics

5.2.1 RG Equations

The RG equations we have derived are ($y = -2$):

$$\frac{d\nu}{d\ell} = \left(z - 2 + \frac{\lambda^2}{\nu}\frac{d^2-2}{2d(d+2)}\left(\frac{A}{\nu^2} + \frac{B}{\mu^2}\right)\frac{S_d}{(2\pi)^d}\Lambda^{d-2}\right)\nu, \tag{5.2.1}$$

$$\frac{d\mu}{d\ell} = \left(z - 2 + \frac{\lambda^2}{\mu(\nu+\mu)}\left[\frac{A}{\nu}\frac{(d-1)}{d} + \frac{B}{\mu}\frac{(d-3)}{d}\right]\frac{S_d}{(2\pi)^d}\Lambda^{d-2}\right)\mu, \tag{5.2.2}$$

$$\begin{aligned}\frac{d\lambda}{d\ell} = \Bigg(\eta - d - 1 &- \frac{\lambda^2\Lambda^{d-2}}{d(d+2)}\frac{S_d}{(2\pi)^d}A\left[\frac{1}{\mu\nu(\nu+\mu)} + \frac{\mu}{\nu^2(\nu+\mu)^2}\right] \\ &+ \frac{\lambda^2\Lambda^{d-2}}{d(d+2)}\frac{S_d}{(2\pi)^d}B\left[\frac{1}{2\mu^3} + \frac{1}{\mu\nu(\nu+\mu)} + \frac{\nu}{\mu^2(\nu+\mu)^2}\right]\Bigg)\lambda,\end{aligned} \tag{5.2.3}$$

$$\frac{dA}{d\ell} = \left(-d - z + 2\xi - 2 + \frac{\lambda^2}{A}\frac{d^2-2}{2d(d+2)}\left[\frac{A^2}{\nu^3} + \frac{B^2}{\mu^3}\right]\frac{S_d}{(2\pi)^d}\Lambda^{d-2}\right)A, \tag{5.2.4}$$

$$\frac{dB}{d\ell} = \left(-d - z + 2\xi - 2 + \frac{\lambda^2}{B}\frac{d-2}{2d}\left[\frac{2AB}{\nu\mu(\nu+\mu)}\right]\frac{S_d}{(2\pi)^d}\Lambda^{d-2}\right)B. \tag{5.2.5}$$

In the description of randomly forced MHD equations we have mentioned in the previous chapters that the flow is characterized by a set of dimensionless couplings.

If we consider the above equations to order λ^2 then we can identify three such effective couplings:

$$x^2 = \frac{\lambda^2 A}{\nu^3}, \quad y^2 = \frac{\lambda^2 B}{\mu^3}, \quad z^2 = \frac{\nu}{\mu}. \tag{5.2.6}$$

These quantities represent the effective Reynolds number, magnetic Reynolds number and the Prandtl number, which measures relatives strength of dissipative forces. Without loss of generality the cut-off has been set to unity, $\Lambda = 1$. For the moment let us consider the equations :

$$\frac{d\nu}{d\ell} = (z-2)\,\nu + \mathscr{A}_d x^2 \nu + \mathscr{A}_d \frac{y^2}{z^2}\nu \tag{5.2.7}$$

$$\frac{d\mu}{d\ell} = (z-2)\,\mu + \frac{z^4 x^2}{z^2+1}\left[\frac{(d-1)}{d}\frac{S_d}{(2\pi)^d}\right]\mu + \frac{y^2}{z^2+1}\left[\frac{(d-3)}{d}\frac{S_d}{(2\pi)^d}\right]\mu, \tag{5.2.8}$$

$$\frac{d\lambda}{d\ell} = (\eta - d - 1)\,\lambda, \tag{5.2.9}$$

$$\frac{dA}{d\ell} = (-d - z + 2\xi - 2)\,A + \mathscr{A}_d x^2 A + \mathscr{A}_d \frac{z^6 y^4}{x^2} A, \tag{5.2.10}$$

$$\frac{dB}{d\ell} = (-d - z + 2\xi - 2)\,B + \frac{x^2 z^4}{z^2+1}\left[\frac{d-2}{d}\frac{S_d}{(2\pi)^d}\right]B. \tag{5.2.11}$$

Note, the choice of $y = -2$ once again corresponds to a fluid near equilibrium, so the fluctuation-dissipation theorem must hold once again. Consider re-writing the equations in terms of the reduced variables:

$$\frac{dx}{d\ell} = \frac{1}{2}x\,(2-d) - \mathscr{A}_d x^3 + \frac{1}{2}\mathscr{A}_d x\left(\frac{z^6 y^4}{x^2} - 3\frac{y^2}{z^2}\right), \tag{5.2.12}$$

$$\frac{dy}{d\ell} = \frac{1}{2}y\,(2-d) + \frac{y x^2 z^4}{z^2+1}\mathscr{B}_d - \frac{3}{2}\frac{y^3}{z^2+1}\mathscr{C}_d, \tag{5.2.13}$$

$$\frac{dz}{d\ell} = \frac{1}{2}\mathscr{A}_d x^2 z + \frac{1}{2}\mathscr{A}_d \frac{y^2}{z^2} z - \frac{1}{2}\frac{z^5 x^2}{z^2+1}\mathscr{E}_d - \frac{1}{2}\frac{z y^2}{z^2+1}\mathscr{C}_d, \tag{5.2.14}$$

where we have defined the following quantities:

$$\mathscr{A}_d = \frac{d^2-2}{2d\,(d+2)}\frac{S_d}{(2\pi)^d}, \quad \mathscr{B}_d = \left(\frac{2-4d}{2d}\right)\frac{S_d}{(2\pi)^d}, \quad \mathscr{C}_d = \frac{(d-3)}{d}\frac{S_d}{(2\pi)^d}, \quad \mathscr{E}_d = \frac{(d-1)}{d}\frac{S_d}{(2\pi)^d}. \tag{5.2.15}$$

Now we can investigate a number of different regimes.

5.2.1.1 Kinetic Regime

Consider a model where the stochastic noise enters only the NS equation. It would feed into the induction equation through the non-linear term in the NS equation which couples it to the magnetic field. Such a regime would correspond to a choice when $B = 0$. Using the equations for the reduced couplings, the so called kinetic regime, would be described by a set of equations:

$$\frac{dx}{d\ell} = \frac{1}{2}x\,(2-d) - \mathscr{A}_d x^3, \tag{5.2.16}$$

$$\frac{dz}{d\ell} = \frac{1}{2}\mathscr{A}_d x^2 z - \frac{1}{2}\frac{z^5 x^2}{z^2+1}\mathscr{E}_d. \tag{5.2.17}$$

We have previously solved the equation for x. The solution reads:

$$x\,(\ell) = \frac{x_0 e^{1/2\varepsilon\ell}}{\left(1 + 2\mathscr{A}_d x_0^2 \frac{(e^{\varepsilon\ell}-1)}{\varepsilon}\right)^{1/2}}. \tag{5.2.18}$$

The equation for z is separable which allows us to write:

$$\int_{z(\ell_0)}^{z(\ell)} \frac{\left(z^2+1\right)dz}{\mathscr{A}_d z^3 + \mathscr{A}_d z - \mathscr{E}_d z^5} = \frac{1}{2}\int_{\ell_0}^{\ell} \frac{x_0 e^{\varepsilon\ell} d\ell}{\left(1 + 2\mathscr{A}_d x_0^2 \frac{(e^{\varepsilon\ell}-1)}{\varepsilon}\right)}. \tag{5.2.19}$$

The RHS admits a compact form upon integration:

$$\frac{1}{2}\int_{\ell_0}^{\ell} \frac{x_0 e^{\varepsilon\ell} d\ell}{\left(1 + 2\mathscr{A}_d x_0^2 \frac{(e^{\varepsilon\ell}-1)}{\varepsilon}\right)} = \frac{1}{4\mathscr{A}_d}\ln\left(1 + 2x_0^2\mathscr{A}_d\,|\varepsilon|^{-1} - 2x_0^2\mathscr{A}_d\,|\varepsilon|^{-1}e^{-|\varepsilon|\ell}\right). \tag{5.2.20}$$

Solution to the LHS is more involved. Although it admits an exact analytic solution, the resultant equation does not permit a simple rearrangement into a solution of the form $z = f\,(\ell)$, where f is some function. For that reason we will not attempt to write the explicit form here. The fixed point value of z can be readily deduced directly from the differential RG equation:

$$\left.\frac{dz}{d\ell}\right|_{z_*} = 0, \quad \text{s.t.} \quad 0 = z_*^4 - \left(\frac{\mathscr{A}_d}{\mathscr{E}_d}\right)z_*^2 - \left(\frac{\mathscr{A}_d}{\mathscr{E}_d}\right). \tag{5.2.21}$$

Recall that because we are looking at the regime where kinetic fluctuations occur in the system near equilibrium the fluctuation dissipation theorem must hold. In that case we have that $x\,(\ell)$ goes to zero as $\lambda\,(\ell)$. This means that in the macroscopic limit, as $\ell \to \infty$, the non-linear terms are suppressed and the resultant equations are linear. In this case, both the induction equation and the NS equation are linear. We can safely ignore all the corrections proportional to λ. For that reason the exclusion of the corrections of λ^3 and higher order terms is justified within our framework.

We can make a particular choice for the dynamic exponent consistent with the fluctuation dissipation theorem:

$$z(\ell) = 2 - \mathscr{A}_d x^2(\ell). \tag{5.2.22}$$

While under this particular choice the viscosity and the noise covariance remain at their initial value, the magnetic diffusivity behaves as follows:

$$\frac{1}{\mu}\frac{d\mu}{d\ell} = \left(-\mathscr{A}_d + \mathscr{E}_d \frac{z^4}{z^2+1}\right) x^2(\ell), \tag{5.2.23}$$

This allows as to deduce the limiting value of μ either by a direct computation:

$$\ln\left(\frac{\mu(\ell)}{\mu(\ell_0)}\right) = \int_{\ell_0}^{\ell} \left(-\mathscr{A}_d + \mathscr{E}_d \frac{z^4(\ell')}{z^2(\ell')+1}\right) x^2(\ell')\, d\ell', \tag{5.2.24}$$

or by noting that once we know the limit of the Prandtl number and the fact that viscosity does not vary along the RG trajectory:

$$\lim_{\ell\to\infty} \mu(\ell) = \frac{\nu(\ell_0)}{z_*^2}. \tag{5.2.25}$$

All of the above considerations are valid for $d \geq 2$. The marginal terms which are generated under the RG transformation are suppressed and, therefore, can be safely dropped. Note that the limiting values depend on the dimensionality of the problem as one can see from the phase-space diagram for $d = 2$ and $d = 3$ (Fig. 5.7).

The kinetically driven model has a stable fixed point at $d = 2$. In this case the effective Reynolds number decays slowly enough, namely as $\ell^{-1/2}$, which permits the RG trajectory to terminate at $(z_*; 0)$. In $d = 3$ the effective Reynolds number dominates over the Prandtl number and therefore the limiting value of $z(\ell)$ strongly depends on the initial values of these parameters $(z(\ell_0); x(\ell_0))$.

The macroscopic behavior of the kinetically driven MHD equations is described by two linear equations:

$$\frac{\partial P_i}{\partial t} = \nabla^2\left(\gamma_+^{\text{eff}} P_i + \gamma_-^{\text{eff}} Q_i\right) + \xi_i, \tag{5.2.26}$$

$$\frac{\partial Q_i}{\partial t} = \nabla^2\left(\gamma_-^{\text{eff}} P_i + \gamma_+^{\text{eff}} Q_i\right) + \xi_i, \tag{5.2.27}$$

where

$$\gamma_\pm^{\text{eff}} = \begin{cases} \gamma_\pm(z(\ell_0); x(\ell_0)) & \text{for} \quad d > 2, \\ \gamma_\pm(z_*) & \text{for} \quad d = 2. \end{cases} \tag{5.2.28}$$

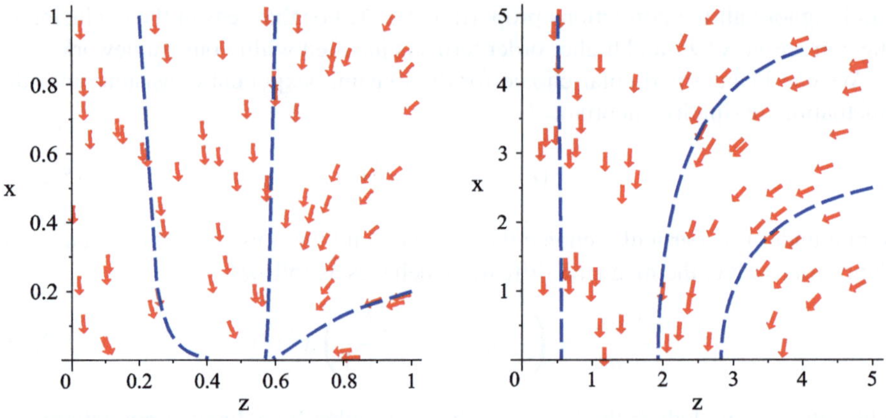

Fig. 5.7 The qualitative difference between $d = 2$ and $d = 3$ stems from the fact that the higher the dimensionality of the problem the faster is the decay of $x(\ell) \to 0$ as $\ell \to \infty$. Despite the fact that z admits a stable fixed point at $d = 2$ and $d = 3$, the behaviour for small $x(\ell)$ and $z(\ell)$ in $d \geq 3$ is completely dominated by the expoential character of $x(\ell)$

The scaling properties of this model are the same as those for the original Model A.

5.2.1.2 Magnetic Regime

We can study the properties of MHD equations when the noise enters the system through the induction equation. Such a model would correspond to setting $A = 0$. The equations governing the magnetically driven system are:

$$\frac{dy}{d\ell} = \frac{1}{2} y (2-d) - \frac{3}{2} \frac{y^3}{z^2+1} \mathscr{C}_d, \tag{5.2.29}$$

$$\frac{dz}{d\ell} = \frac{1}{2} \mathscr{A}_d \frac{y^2}{z} - \frac{1}{2} \frac{zy^2}{z^2+1} \mathscr{C}_d. \tag{5.2.30}$$

Equations take a particularly simple form in $d = 3$ because $\mathscr{C}_{d=3} = 0$. In turn this allows us to solve them exactly:

$$y(\ell) = y(\ell_0) e^{-1/2\ell}, \tag{5.2.31}$$

$$z(\ell) = \left(z^2(\ell_0) + 2\mathscr{A}_d y(\ell_0) e^{-1/2\ell_0} - 2\mathscr{A}_d y(\ell_0) e^{-1/2\ell} \right)^{1/2}. \tag{5.2.32}$$

The behavior of the solution suggests that the effective coupling goes to zero exponentially fast while $z(\ell)$ goes to a value which is determined by the initial conditions. This is depicted below (Fig. 5.8):

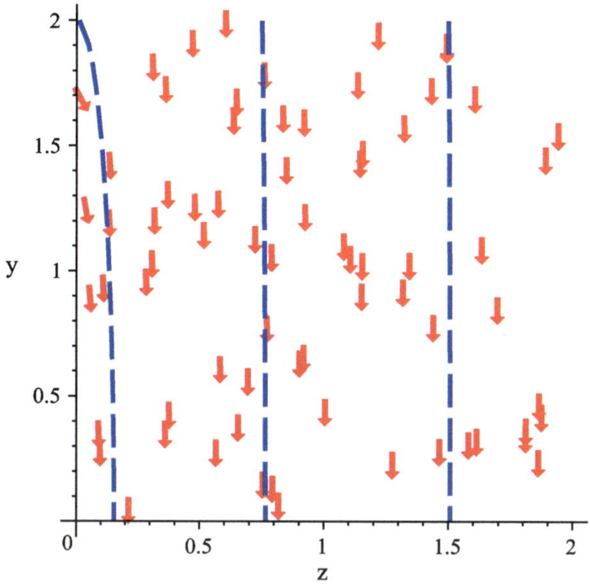

Fig. 5.8 RG trajectories for magnetically driven MHD equations in $d = 3$

From the solutions to the RG equations we observe that there exists a non-trivial limit for z, which strongly depends on the initial conditions:

$$\lim_{\ell\to\infty} z(\ell) = \left(z^2(\ell_0) + 2\mathscr{A}_d y(\ell_0) e^{-1/2\ell_0}\right)^{1/2}. \tag{5.2.33}$$

The application of the fluctuation-dissipation theorem to this model is a non-trivial task. Recall that in previous studies we chose the dynamic exponent in such a manner so that the theorem holds. An important consequence of the theorem was to render the macroscopic theory linear while all of the non-linear terms could be consistently ignored. This is necessary because otherwise we generate terms which are marginal. Higher order analysis is problematic as there are infinitely many such terms in our expansion. Thus, we require some other mechanism which would suppress such terms in a consistent manner. The differential RG equations governing magnetically forced MHD equations are:

$$\frac{d\nu}{d\ell} = (z-2)\nu + \mathscr{A}_d \frac{y^2}{z^2}\nu, \tag{5.2.34}$$

$$\frac{d\mu}{d\ell} = (z-2)\mu, \tag{5.2.35}$$

$$\frac{d\lambda}{d\ell} = (\eta - d - 1)\lambda, \tag{5.2.36}$$

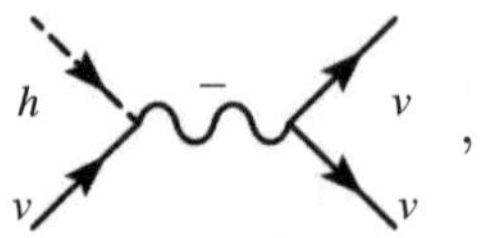

Fig. 5.9 A marginal term in the NS RG calculation

$$\frac{dB}{d\ell} = (-d - z + 2\xi - 2)\, B. \tag{5.2.37}$$

By explicitly solving the equations for the reduced couplings we have obtained:

$$\lambda^2(\ell) = \frac{\lambda^2(\ell_0)\, B(\ell_0)}{\mu^3(\ell_0)} \mu^3(\ell)\, B^{-1}(\ell)\, e^{-\ell}. \tag{5.2.38}$$

In order to drive all the marginal terms to zero we have to make a choice for the scaling exponents such that:

$$z(\ell) = 2, \quad \xi(\ell) = z(\ell) + d/2, \quad \forall \ell. \tag{5.2.39}$$

These exponents remain constant throughout the RG trajectory and $\lambda \to 0$ exponentially fast. While the magnetic noise and magnetic diffusivity remain constant, the kinematic viscosity, ν, tends to a fixed point value:

$$\lim_{\ell\to\infty} \nu(\ell) = \left(\nu^2(\ell_0) + 2\mathscr{A}_d(\ell_0)\,\lambda(\ell_0)\,\mu^{\frac{1}{2}}(\ell_0)\, B^{\frac{1}{2}}(\ell_0)\, e^{-1/2\ell_0}\right)^{1/2}. \tag{5.2.40}$$

The equations governing the macroscopic behavior are linear, with the effective coefficients determined by the above equations. For $d = 2$ the theory is unstable in the sense that both the Prandtl number and the effective magnetic Reynolds number blow up to infinity. This is a consequence of the change of sign in the coefficient, $\mathscr{C}_{d<3} < 0$.

5.2.2 The Problem of Marginal Variables

Having studied both the magnetic and the kinetic regimes it would be natural to consider the cross-over regime when both stirring forces are switched on. The problem with this regime is that it is not clear how one would incorporate the fluctuation dissipation theorem and whether or not it would allow us to consistently suppress terms which are otherwise marginal. In fact the problem of marginal terms is more severe for the MHD equations. While in the turbulent NS we had to worry about a graph such as the one depicted below (Fig. 5.9):

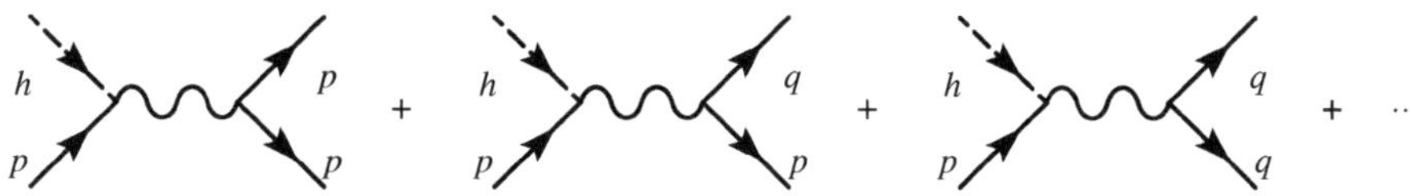

Fig. 5.10 A marginal term in the NS RG calculation

In MHD we have many more such graphs (Fig. 5.10):

Since the scaling argument is the same, the MHD system has marginal terms to all orders in λ, which makes a higher-order analysis problematic. To this end our analysis is limited to situations when the system is near equilibrium. In what follows we discuss possible directions in which the study of MHD turbulence can be taken.

5.2.3 Non-perturbative Methods

5.2.3.1 Galilean Symmetry

Similar to the NS equation, the Galilean transformation is a symmetry of the MHD equations. Consider the following transformation:

$$\mathbf{x}' = \mathbf{x}'(\mathbf{x}, t), \ t' = t'(\mathbf{x}, t), \tag{5.2.41}$$

such that

$$\mathbf{x}' = \mathbf{x} - \mathbf{c}t, \ t' = t, \tag{5.2.42}$$

where $\mathbf{c}$ is a constant. It follows that differential operators transform as:

$$\frac{\partial}{\partial x_i} = \delta_{ij}\frac{\partial}{\partial x_j}, \tag{5.2.43}$$

$$\frac{\partial}{\partial t} = -c_i\frac{\partial}{\partial x_i'} + \frac{\partial}{\partial t'}. \tag{5.2.44}$$

We have already established that the NS equation is invariant under the above transformation given we transform the velocity field as:

$$v_i'\left(\mathbf{x}', t'\right) = v_i\left(\mathbf{x}' + ct', t'\right) - c_i. \tag{5.2.45}$$

In addition we require the magnetic field which enters the NS equation to transform as:

$$b_i'\left(\mathbf{x}', t'\right) = b_i\left(\mathbf{x}' + ct', t'\right). \tag{5.2.46}$$

The kinetic equation is then invariant. Given the above field and operator transformations, the coupling term in the induction equation generates a term:

$$\nabla \times (\mathbf{c} \times \mathbf{b}) = \mathbf{c}\,(\nabla \cdot \mathbf{b}) - \mathbf{b}\,(\nabla \cdot \mathbf{c}) + (\mathbf{b} \cdot \nabla)\,\mathbf{c} - (\mathbf{c} \cdot \nabla)\,\mathbf{b}. \tag{5.2.47}$$

The first term vanishes because of the incompressibility condition. The second and the third term are zero because the differential operator acts on a constant **c**. The last term survives but cancels the contribution which comes from the temporal differential operator under the transformation:

$$\frac{\partial b_i}{\partial t} = -c_i \frac{\partial b'_k}{\partial x'_i} + \frac{\partial b'_k}{\partial t'}. \tag{5.2.48}$$

Therefore, the MHD equation are invariant under the Galilean transformation.

5.2.3.2 Ward-Takahashi Identities

The Galilean transformation is a continuous symmetry. It therefore follows that we can consider infinitesimal transformations by treating **c** as an infinitesimal quantity. As a consequence of the symmetry, the MHD action must remain invariant and infinitesimal changes induced into the action by the transformation must vanish.

As we have stressed in the previous discussion of the Galilean transformation, it is instructive to use Reynolds decomposition. In this case the propagator is explicitly Galilean invariant. However it is advantageous, in the context of MHD, to work with Elssaser variables, because the equations take a particularly symmetric form. Recall, the MHD action is:

$$\begin{aligned}\mathscr{S}\,[\mathbf{P}, \mathbf{Q}, \mathbf{h}, \mathbf{n}] = \int d\mathbf{x}\, dt \Bigg[& \frac{1}{2} \int d\mathbf{y}\, h_i\,(\mathbf{x}, t)\, A_{ij}\,(\mathbf{x} - \mathbf{y})\, h_j\,(\mathbf{y}, t) \\ & + \frac{1}{2} \int d\mathbf{y}\, n_i\,(\mathbf{x}, t)\, A_{ij}\,(\mathbf{x} - \mathbf{y})\, n_j\,(\mathbf{y}, t) + \frac{1}{2} \int d\mathbf{y}\, h_i\,(\mathbf{x}, t)\, B_{ij}\,(\mathbf{x} - \mathbf{y})\, n_j\,(\mathbf{y}, t) \\ & - i h_i \left(\frac{\partial P_i}{\partial t} + \lambda \mathrm{P}_{ij} \frac{\partial P_j Q_k}{\partial x_k} - \gamma_+ \nabla^2 P_i - \gamma_+ \nabla^2 Q_i \right) \\ & - i n_i \left(\frac{\partial Q_i}{\partial t} + \lambda \mathrm{P}_{ij} \frac{\partial P_k Q_j}{\partial x_k} - \gamma_+ \nabla^2 Q_i - \gamma_+ \nabla^2 P_i \right) \Bigg].\end{aligned} \tag{5.2.49}$$

Reynolds decomposition breaks down the field variable into an average and a fluctuating part:

$$\mathbf{P} = \mathbf{P}_{av} + \mathbf{p}, \tag{5.2.50}$$

$$\mathbf{Q} = \mathbf{Q}_{av} + \mathbf{q}. \tag{5.2.51}$$

Thus, the generating functional takes the form:

$$\mathscr{Z}[\mathbf{J},\mathbf{G},\mathbf{j},\mathbf{g}] = \int d\mu \exp\left[-\mathscr{S} + \int d\mathbf{x}\, dt \left(\mathbf{J}\cdot(\mathbf{P}_{av}+\mathbf{p}) + \mathbf{j}\cdot h + \mathbf{G}\cdot(\mathbf{Q}_{av}+\mathbf{q}) + \mathbf{g}\cdot n\right)\right], \tag{5.2.52}$$

where we have used a shorthand notation:

$$d\mu = [\mathscr{D}\mathbf{p}][\mathscr{D}\mathbf{q}][\mathscr{D}\mathbf{h}][\mathscr{D}\mathbf{n}]. \tag{5.2.53}$$

Under the co-ordinate transformations:

$$\mathbf{x}' = \mathbf{x} - \mathbf{c}t, \qquad t' = t, \tag{5.2.54}$$

all fields transform according to the rule:

$$\mathbf{F}'_{av} = \mathbf{F}_{av} - \mathbf{c}, \tag{5.2.55}$$

$$\mathbf{f}'(\mathbf{x}',t') = \mathbf{f}(\mathbf{x}',t') + t'\delta\mathbf{c}\cdot\nabla\mathbf{f}(\mathbf{x}',t'). \tag{5.2.56}$$

Under the Galilean transformation the generating functional for the correlation functions transforms as:

$$\mathscr{Z} \longrightarrow \mathscr{Z} + \delta\mathscr{Z}. \tag{5.2.57}$$

Because it a symmetry transformation we require:

$$\delta\mathscr{Z} = 0. \tag{5.2.58}$$

This condition translates into an integro-differential equation:

$$0 = \int d\mathbf{x}\, dt \left[J_i(\mathbf{x},t)\left(t\delta\mathbf{c}\cdot\nabla\frac{\delta}{\delta J_i(\mathbf{x},t)} - \delta c_i\right) + G_i(\mathbf{x},t)\left(t\delta\mathbf{c}\cdot\nabla\frac{\delta}{\delta G_i(\mathbf{x},t)} - \delta c_i\right)\right.$$
$$\left. + j_i(\mathbf{x},t)\, t\delta\mathbf{c}\cdot\nabla\frac{\delta}{\delta j_i(\mathbf{x},t)} + g_i(\mathbf{x},t)\, t\delta\mathbf{c}\cdot\nabla\frac{\delta}{\delta g_i(\mathbf{x},t)}\right]. \tag{5.2.59}$$

The most convenient form to explore properties between correlation functions is found by making a Legendre transformation, which we define below:

$$\Gamma\left[\mathbf{p}^{cl},\mathbf{q}^{cl},h^{cl},n^{cl}\right] = -\mathscr{F}[\mathbf{J},\mathbf{G},\mathbf{j},\mathbf{g}] + \int d\mathbf{x}\, dt \left(\mathbf{J}\cdot\mathbf{p}^{cl} + \mathbf{G}\cdot\mathbf{q}^{cl} + \mathbf{j}\cdot h^{cl} + \mathbf{g}\cdot n^{cl}\right), \tag{5.2.60}$$

where the fields are defined as

$$\frac{\delta\mathscr{F}}{\delta J_i} = p_i^{cl}, \quad \frac{\delta\mathscr{F}}{\delta G_i} = q_i^{cl}, \quad \frac{\delta\mathscr{F}}{\delta j_i} = h_i^{cl}, \quad \frac{\delta\mathscr{F}}{\delta g_i} = n_i^{cl}. \tag{5.2.61}$$

As in the usual Legendre transform, when we re-express information in terms of derivatives, here we re-express the theory in terms of average realizations of the field. The equation which relates correlation function is then transformed into a statement relating various loops:

$$0=\int d\mathbf{x}\,dt\left[\frac{\delta\Gamma}{\delta p_i^{cl}(\mathbf{x},t)}\left(t\delta\mathbf{c}\cdot\nabla p_i^{cl}(\mathbf{x},t)-\delta c_i\right)+\frac{\delta\Gamma}{\delta q_i^{cl}(\mathbf{x},t)}\left(t\delta\mathbf{c}\cdot\nabla q_i^{cl}(\mathbf{x},t)-\delta c_i\right)\right.$$
$$\left.+\frac{\delta\Gamma}{\delta h_i^{cl}(\mathbf{x},t)}t\delta\mathbf{c}\cdot\nabla h_i^{cl}(\mathbf{x},t)+\frac{\delta\Gamma}{\delta n_i^{cl}(\mathbf{x},t)}t\delta\mathbf{c}\cdot\nabla n_i^{cl}(\mathbf{x},t)\right]. \tag{5.2.62}$$

By succesively taking derivatives with respect to classical fields we form an infinite hierarchy of equations relating loop diagrams. For instance we find a relationship which relates the propagator loop with the three-point vertex:

$$0=t'\delta\mathbf{c}\cdot\nabla_y\frac{\delta^2\Gamma}{\delta h_n^{cl}(\mathbf{z},t'')\,\delta p_s^{cl}(\mathbf{y},t')}+t''\delta\mathbf{c}\cdot\nabla_z\frac{\delta^2\Gamma}{\delta h_n^{cl}(\mathbf{z},t'')\,\delta p_s^{cl}(\mathbf{y},t')}$$
$$+\delta c_i\int d\mathbf{x}\,dt\left(\frac{\delta^3\Gamma}{\delta h_n^{cl}(\mathbf{z},t'')\,\delta p_s^{cl}(\mathbf{y},t')\,\delta q_i^{cl}(\mathbf{x},t)}+\frac{\delta^3\Gamma}{\delta h_n^{cl}(\mathbf{z},t'')\,\delta p_i^{cl}(\mathbf{x},t)\,\delta p_s^{cl}(\mathbf{y},t')}\right). \tag{5.2.63}$$

This is a non-perturbative technique and is, therefore, valid to all orders in the perturbation theory. It is instructive to check the validity of this results to zero order in the number of loops. To zero order the last term is zero, since there is no term of this form present in the action. In addition, the above relation is more instructive in Fourier space. First, we make use of translational invariance and approximate the relation to zero order. This would have the form:

$$\delta\mathbf{c}\cdot\left(t''-t'\right)\nabla_y\Gamma^0_{h_n^{cl}p_s^{cl}}\left(0,0;\mathbf{y}-\mathbf{z},t'-t''\right)$$
$$=\delta c_i\int d\mathbf{x}\,dt\,\Gamma^0_{h_n^{cl}p_s^{cl}q_i^{cl}}\left(\mathbf{0},0;\mathbf{y}-\mathbf{z},t'-t'';\mathbf{x}-\mathbf{z},t-t''\right) \tag{5.2.64}$$

By transforming the above relationship into Fourier space, we obtain:

$$-k_i\frac{\partial}{\partial\omega}\Gamma^0_{h_n^{cl}p_s^{cl}}(\mathbf{k},\omega;-\mathbf{k},-\omega)=\Gamma^0_{h_n^{cl}p_s^{cl}q_i^{cl}}(\mathbf{0},0;\mathbf{k},\omega;-\mathbf{k},-\omega). \tag{5.2.65}$$

By substituting the functions explicitly the identity clearly holds.

$$-k_i\frac{\partial}{\partial\omega}\left(i\omega-i\mathbf{O}\cdot\mathbf{k}+\gamma_+k^2\right)\mathrm{P}_{sn}=-ik_i\mathrm{P}_{sn}. \tag{5.2.66}$$

This is, in fact, a rather trivial modification of the NS result. The vector $\mathbf{O}$ is a constant drift field, which enters the equations as a consequence of the Reynolds decomposition. However, as we increase the number of loops, the verification of the above identities becomes a non-trivial task. Consider corrections up to one loop. Then we can write:

$$\Gamma^{1}_{h_n^{cl} p_s^{cl} q_i^{cl}} (\mathbf{0}, 0; \mathbf{k}, \omega; -\mathbf{k}, -\omega) = -\, i k_i \mathrm{P}_{sn} + \Lambda_1 (\mathbf{k}, \omega), \tag{5.2.67}$$

$$\Gamma^{1}_{h_n^{cl} p_s^{cl} p_i^{cl}} (\mathbf{0}, 0; \mathbf{k}, \omega; -\mathbf{k}, -\omega) = \Lambda_2 (\mathbf{k}, \omega). \tag{5.2.68}$$

It follows that Λ_1, which is an additive correction from the 1-loop RG, does not have to vanish, like in the NS case, and we have the following identity:

$$\Lambda_1 (\gamma_+, \gamma_-, A, B) = -\Lambda_2 (\gamma_+, \gamma_-, A, B). \tag{5.2.69}$$

This is a consequence of the fact that other loop corrections enter the equations. In our analysis, despite the fact that we have computed non-trivial corrections to the vertex, we have focused solely on the regime where non-linear terms are irrelevant in the macroscopic limit and therefore did not affect our analysis. It remains to be shown that the correction to the three-point vertex, computed in our calculation, is consistent with Ward-Takahashi identities. Such investigations can form the basis for future studies of the problem.

5.2.4 Future Research

There are many interesting problems associated with the functional integral formulation of stochastic differential equations such as RG analysis, non-locality of the action and symmetry considerations which can be extended to MHD from the existing literature on the NS equation. We will outline them in the next section to demonstrate potential directions in which future research can be taken.

5.2.4.1 BRS Symmetry and Gauge Fixing in the Functional Integral Formulation

Symmetry considerations are extremely useful, as they often allow us to simplify a problem in question. As we have just seen, the Galilean transformation, for example, provides relationships between various vertex function which could not be established otherwise. However, there are cases when a continuous symmetry can complicate matters at hand.

We first consider a simple problem in combinatorics when we wish to count a number of non-equivalent realizations of placing n identical/indistinguishable objects into m boxes ($n \leq m$). First we count how many ways there are of placing n objects into m boxes, this is given by:

$$C = \frac{m!}{(m-n)!}. \tag{5.2.70}$$

Since all n are identical we are in fact over-counting by a factor of $n!$, which is a number of ways we can permute n objects. So we conclude that to count all non-equivalent realizations we have to compute:

$$\frac{m!}{n!\,(m-n)!}. \tag{5.2.71}$$

Note, we have divided through by the number of identical realization.

Now, let us look at a slightly different example: consider a $2D$ action in ordinary calculus which is a function of two variables:

$$S = S\,(x, y)\,. \tag{5.2.72}$$

Suppose the action is invariant under rotations and, therefore, is a function of r only. We can think of the action as follows: for every r, S specifies a physical system, while a rotation would relate two equivalent physical systems. Since we are interested in counting all non-equivalent physical states, in this case we are again over-counting. In the context of statistical mechanics, we would be looking at an expression of the form:

$$Z = \int dx\, dy\, e^{-S(x,y)}. \tag{5.2.73}$$

Due to the rotational symmetry of the action we can write:

$$Z = \int_0^{2\pi} d\theta \int_0^{\infty} r dr\, e^{-S(r)}. \tag{5.2.74}$$

To count all non-equivalent realization we have to factor out 2π, which can be thought of as the volume of the symmetry group. Thus, we have:

$$Z_{\text{non-equiv}} = (2\pi)^{-1}\, Z. \tag{5.2.75}$$

Dividing through by either a number of equivalent realizations or a volume of the symmetry group works as long as this quantities are finite. This method fails if we are to consider problems with translational invariance or Galilean invariance, for that matter, since the volume of these symmetry groups is infinite.

The method which does not have an explicit division by the symmetry group volume requires us to introduce the formalism of Grassmann variables [8]. It is appropriate to introduce this formalism here with a number of results because it will form the basis for other discussions in the next sections.

Grassmann algebra. A Grassmann algebra $\mathscr{A}$ is an algebra constructed from a set of generators θ_i and their anti-commuting products:

$$\theta_i\theta_j + \theta_j\theta_i = 0 \quad \forall \quad i,\ j. \tag{5.2.76}$$

The consequence of the anti-commuting property of Grassmann variables is that an arbitrary analytic function defined on this algebra is a first degree polynomial in each element. For example:

$$f\left(\theta_i,\theta_j\right)=a_0+a_1\theta_i+a_2\theta_j+a_3\theta_i\theta_j. \tag{5.2.77}$$

Differentiation is defined in the same way as for complex variables, with the exception that the order of differentiation matters. Using the general function defined above we observe:

$$\frac{\partial^2 f}{\partial\theta_i\partial\theta_j}=-\frac{\partial^2 f}{\partial\theta_j\partial\theta_i}. \tag{5.2.78}$$

While differentiation is analogous to the definition with ordinary variables, integration over Grassmann variables does not correspond to Riemann integrals. Instead, the operation of integration is defined so that it resembles the fundamental property of an ordinary integral over the exact differentials of functions which vanish at infinity. As a consequence we have:

$$\int d\theta=0, \tag{5.2.79}$$

because 1 is an exact differential of θ. Further, one defines:

$$\int\theta d\theta=1. \tag{5.2.80}$$

Since θ is not a derivative of $1/2\theta^2$ in Grasmann algebra, the usual notion of exact differentials fails. Instead, the integration can be viewed as a linear mapping onto the positive real numbers. This definition is made precise by following the standard convention.

The above results are now sufficient to appreciate some of the more complicated computations. For our purposes we are interested in Gaussian integrals defined on the Grassmann algebra $\mathscr{A}$. It is instructive to appreciate how to change variables in Grassmann integrals. Consider:

$$\int d\theta f\left(\theta\right)=\int d\theta\left(a_0+a_1\theta\right). \tag{5.2.81}$$

Let us perform a change of variables:

$$\theta=b_0+b_1\eta. \tag{5.2.82}$$

The value of the integral remains unchanged, however a Jacobian of the transformation must be introduced:

$$\int d\theta f\left(\theta\right)\to b_1^{-1}\int d\eta f\left(b_0+b_1\eta\right)=a_1. \tag{5.2.83}$$

The result is very different from ordinary calculus where the Jacobain determinant would have been b_1.

Let us go back to the problem of degenerate states. In the context of the randomly stirred NS equation, the first realization of the problem caused by the Galilean transformation is due to [10]. Here, we would like to go through the points made in this chapter and propose further investigation of the matter in the context of MHD.

Galilean transformation is a continuous symmetry which relates velocity fields when viewed from different frames of reference. The functional integral

$$\mathscr{Z} = \int \mathscr{D}[\mathbf{V}]\,\mathscr{D}[\sigma]\exp(-S[\mathbf{V},\sigma]), \tag{5.2.84}$$

where the action is defined as:

$$S[\mathbf{V},\sigma] = \int d\mathbf{x}\int dt\left[\frac{1}{2}\int d\mathbf{y}\sigma_i(\mathbf{x},t)\,D_{ij}(\mathbf{x}-\mathbf{y})\,\sigma_j(\mathbf{y},t)\right.$$
$$\left. - i\sigma_k\left(\frac{\partial V_k}{\partial t} + \lambda_0 P_{kj}(\nabla)\frac{\partial\left(V_i V_j\right)}{\partial x_i} - \nu_0\nabla^2 V_k\right)\right], \tag{5.2.85}$$

is invariant under such transformation and, therefore, over counts physical states by an infinite factor. This is similar to over-counting of the orbits in our previous example, however, here this over-counting is not finite and cannot be factored in a simple manner. As a result one seeks first to choose a representative of the physical state by fixing an inertial frame. Formally this is referred to as fixing the gauge. For instance one can choose to fix the drift term, which in Fourier space would correspond to the zero mode component of the velocity field:

$$\mathscr{Z} = \int \mathscr{D}[\mathbf{V}]\,\mathscr{D}[\sigma]\,\delta[\mathbf{V}_0 - \mathbf{b}]\exp(-S[\mathbf{V},\sigma]), \tag{5.2.86}$$

This can be formally re-written as follows:

$$\mathscr{Z}' = \int \mathscr{D}[\mathbf{V}]\,\mathscr{D}[\sigma]\int d\mathbf{b}\delta[\mathbf{V}_0 - \mathbf{b}]\exp\left(-S[\mathbf{V},\sigma] - \frac{\mathbf{b}\cdot\mathbf{b}}{2\xi}\right). \tag{5.2.87}$$

Both choices are equivalent. We have suppressed all field independent prefactors, but motivated by our previous discussion on Jacobian determinants in ordinary calculus, in the limit of $\xi \to 0$, one indeed recovers the δ-function. Thus, we end up with the expression:

$$\mathscr{Z} = \int \mathscr{D}[\mathbf{V}]\,\mathscr{D}[\sigma]\exp\left(-S[\mathbf{V},\sigma] - \frac{\mathbf{V}_0\cdot\mathbf{V}_0}{2\xi}\right). \tag{5.2.88}$$

This can also be viewed as follows, instead of giving a sharp cut-off in the form of the δ-function, we assign a Gaussian weight to different inertial frames, which is a lot smoother. Such tricks also take place in a slightly different context [11], to which we will come back. To this end we have fixed the gauge, however the Galilean transformation is explicitly broken by the last term. There exists a mathematical construct which allows to identify another symmetry of the system, which is not physical. Consider multiplying the generating functional by some field independent factor:

$$\mathscr{Z} = \int \mathscr{D}[\mathbf{V}]\mathscr{D}[\sigma]\int d\theta d\bar{\theta}\exp\left(-S[\mathbf{V},\sigma] - \frac{\mathbf{V}_0\cdot\mathbf{V}_0}{2\xi} + i\bar{\theta}\cdot\theta\right) \quad (5.2.89)$$

where θ is a Grassmann vector and $\bar{\theta}$ is a complex conjugate. This may seem like an ad hoc step, however, the rational behind it is to bring the determinant, which in our case is **1**, into the action. For if we have made a more complicated choice to fix the gauge, we would have inserted a factor of unity of the form:

$$\rightarrow \det\left(\frac{\delta F_i(\mathbf{x})}{\delta V_j(\mathbf{y})}\right)\delta[F[\mathbf{V}]]. \quad (5.2.90)$$

This is where the Grassmann variables become very handy. We have seen that the change of variables in Grassmann integrals is different to ordinary integration rules, in particular the multi-dimensional Gaussian integral then reads:

$$\begin{aligned} I(\mathbf{a}) &= \int d\theta_1 d\bar{\theta}_1 \dots d\theta_n d\bar{\theta}_n \exp\left(\sum_{i,j}\bar{\theta}_i a_{ij}\theta_j\right) \\ &= \det \mathbf{a}. \end{aligned} \quad (5.2.91)$$

Introduction of Grassmann fields simply lifts the Jacobian of the transformation into the action. Finally, the resultant action:

$$\tilde{S} = S[\mathbf{V},\sigma] + \frac{\mathbf{V}_0\cdot\mathbf{V}_0}{2\xi} - i\bar{\theta}\cdot\theta, \quad (5.2.92)$$

posses a new symmetry:

$$\mathbf{V}_0' = \mathbf{V}_0 - c\zeta\left(\bar{\theta}+\theta\right), \quad (5.2.93)$$

$$\theta' = \theta - \frac{i}{\xi}\mathbf{V}_0 c\zeta, \quad (5.2.94)$$

$$\bar{\theta}' = \bar{\theta} + \frac{i}{\xi}\mathbf{V}_0 c\zeta, \quad (5.2.95)$$

while the non-zero modes of the remaining fields transform as before, with the boost velocity substituted $\mathbf{c} \rightarrow c\zeta\left(\bar{\theta}+\theta\right)$, where c is a constant with dimensions of velocity. This new symmetry is a mathematical artifact, rather than physically meaningful symmetry, so the problem of degenerate physical states does not arise.

Further, one can exploit the newly defined symmetry, known as Becchi-Rouet-Stora (BRS) symmetry [12], to derive Ward-Takahashi identities.

Such considerations clearly can be extended to the MHD equations. It is necessary for a well-defined functional integral formulation of the problem. The relations, which result from symmetry considerations, are analogous to the Slavnov-Taylor identities in QFT [10] and to our knowledge, to date, no such calculations have been carried out for the randomly stirred MHD.

5.2.4.2 Further Extensions

Similar symmetry application can be applied to a number of other field transformations. For instance, the condition of incompressibility can also be treated by using the same formalism of Grassmann variables [11]. Recall, we have implemented this condition through the functional Fourier transform of the δ-function, while one could, using ideas we have talked through in the previous section, treat it through Grassmann variables. As we have mentioned such methods have an advantage over the strict definition of the δ-function, as one has a free parameter which, when non-zero, suppresses fluctuations from the strict incompressibility condition by Gaussian weight. This is analytically better behaved and one chooses such approach when attempts to solve problems numerically.

It would be interesting to use Grassmann fields, known as ghosts [8], in the context of MHD determinant. Direct computation of the functional determinant should be completely equivalent. The former however presents the problem through Feynman diagram and it would be useful to carry out such computations for completeness.

Another important point, which should be kept in mind in order to investigate the MHD system numerically, is the non-local nature of some of the terms in the action [11]. These terms are proportional to ∇^{-2}, which in Fourier space are proportional to k^{-2}. The cost of re-writing such terms in terms of local interaction comes at the cost of an introduction of new fields. The basic idea is as follows. Consider the term of the form:

$$e^{K\nabla^{-2}L}. \tag{5.2.96}$$

One can insert a field independent factor into the action:

$$I = \int [\mathscr{D}\psi]\, e^{-\int \psi(\mathbf{x}_1,t_1)\nabla^2\psi(\mathbf{x}_2,t_2)d\mathbf{x}_1 d\mathbf{x}_2 dt_1 dt_2}. \tag{5.2.97}$$

where ψ is a non-physical, auxiliary field. Then, through an introduction of the following transformation:

$$\hat{\psi} := \psi + \lambda^{-1}\nabla^{-2}K + \frac{\lambda}{2}\nabla^{-2}L, \tag{5.2.98}$$

the action is transformed into the following form:

$$K\nabla^{-2}L - \hat{\psi}\nabla^2\hat{\psi} \propto -\lambda^{-2}K\nabla^{-2}K - \frac{1}{4}\lambda^2 L\nabla^{-2}L + \text{local terms}. \qquad (5.2.99)$$

The remaining two terms are still non-local but they can be transformed in a similar manner, via introduction of other auxiliary fields. In effect this is like doing a Gaussian integral with a source backwards.

Such manipulations are important if one uses Wegner-Houghton integro-differential equation to carry out RG analysis for example [11, 13]. Since it cannot be evaluated exactly, schemes like a derivative expansion can be implemented to compute lowest order corrections. We note that we are not aware of any attempts to study randomly forced MHD equations via integro-differential RG equations. As we have seen in our study of the Coulomb gas, integro-differential RG equations, which depict infinitesimal change in the action, as a result of the RG, in a form of a differential equation can offer a simpler analysis. For instance, the irreducible differential formulation of Wilson RG, has fewer graphs. All of the disconnected and reducible Feynman diagrams are not present in the formalism. This can be advantageous for the study of turbulence because this means reducible diagrams, will not be generated at all. The flow of the action at the intermediate stages of the RG should be smooth and the presence of marginal terms would be manifested by the initial discontinuity of the flow i.e. at $\ell = 0$.

Application of Wegner-Houghton formalism has been carried for the Navier-Stokes turbulence [11]. As a result the above transformation into local interactions were carried out to implement a derivative expansion. Such calculations are very cumbersome as the number of graphs is very large. In particular, the effect of the discussed transformation increases the the order of the derivatives acting on the physical fields, as a result the derivative expansion has to be carried out to higher order. The authors report that they have implemented computer algorithms to correctly account for all possible diagrams. Unlike the irreducible differential formulation of Wilson RG, Wegner-Houghton equation generates both reducible and irreducible graphs, which makes it a difficult method to implement to models such as turbulence.

We have not been able to track the treatment of marginal terms. However, regardless of that, it is apparent that similar analysis of the MHD equations is very cumbersome. In that sense the irreducible differential formulation of Wilson RG should be the method of choice as it offers computational advantages. We are not aware of any calculations of that sort for the MHD and it would be an interesting to investigate this matter further.

References

1. D. Forster, D.R. Nelson, M.J. Stephen, Large distance and long time properties of a randomly stirred fluid. Phys. Rev. A **16**, 732 (1977)
2. V. Yakhot, S.A. Orszag, Renormalization group analysis of turbulence. Phys. Rev. Lett. **57**, 1722 (1986)

3. G.L. Eyink, The renormalization group method in statistical hydrodynamics. Phys. Fluids **6**, 3063 (1994)
4. W.D. McComb, Galilean invariance and vertex renormalization in turbulence theory. Phys. Rev. E **71**, 037301 (2005)
5. A. Berera, D. Hochberg, Galilean invariance and homogeneous anisotropic randomly stirred flows. Phys. Rev. E **72**, 057301 (2005)
6. J.C. Ward, Phys. Rev. **78**, 1824 (1950)
7. Y. Takahashi, Nuovo Cimento **6**, 370 (1957)
8. J. Zinn-Justin, *Lecture note in Physics* (Springer, Berlin, 1975)
9. R.K.P. Zia, E.F. Redish, S.R. McKay, Making sense of the legendre transform, arXiv : 0806.1147 (2008)
10. A. Berera, D. Hochberg, Gauge symmetry and Slavnov-Taylor identities for randomly stirred fluids. Phys. Rev. Lett. **99**, 254501 (2007)
11. D. Homeier, G. Münster, Renormalization group analysis of turbulent hydrodynamics, arXiv : 1012.0461 (2010)
12. C. Becchi, A. Rouet, R. Stora, Renormalization of the abelian Higgs-Kibble model. Commun. Math. Phys. **42** (1975)
13. F.J. Wegner, A. Houghton, Renormalization group equations for critical phenomena. Phys. Rev. A **8**, 401 (1973)

Chapter 6
Conclusion

6.1 Coulomb Gas

In the context of equilibrium statistical mechanics we have successfully applied an irreducible RG method to the d-dimensional Coulomb gas. The lattice model serves as a convenient starting point for the functional integral formulation of the problem. By studying the properties of differential RG equations we have established, in accordance with previous literature, that in $d = 2$ the system undergoes the celebrated Kosterlitz-Thouless phase transition. While for dimensions $d \geq 3$, there are no physically acceptable fixed points the RG equations can be used to investigate properties of other physical quantities, such as the Debye length. In particular, by drawing analogies with PIC simulations and the underlying lattice model, RG method provides a quantitative tool to study the effects of lattice's linear scales, such as lattice spacing and the system size, on the physical quantities. As an example we have considered the behavior of the Debye length under RG, where such linear scales are incorporated into the differential equation. Eventually one requires an explicit algorithm which would closely resemble the lattice structure formulation in order to be able to make more quantitative predictions which could be directly checked via simulation. This could be a subject for further investigations.

6.2 Randomly Stirred Fluids

Our investigation of non-equilibrium charged systems led us to stochastic partial differential equations which govern extraterrestrial plasmas, the MHD equations. In order to facilitate the RG group analysis we have chosen to work in the functional integral formulation. This introduced a new feature, which was absent in the Coulomb gas problem, namely the Jacobian determinant. Our analysis has shown that by exponentiating the Jacobian determinant and considering term by term in the resultant expansion, the resultant contribution is a field independent constant and therefore can

E. Barkhudarov, *Renormalization Group Analysis of Equilibrium and Non-equilibrium Charged Systems*, Springer Theses,
DOI: 10.1007/978-3-319-06154-2_6,

be dropped. Alternatively, one could use Grassmann fields to analyze the determinant. To our knowledge such considerations have not been treated in the MHD literature and could serve as an interesting problem for future investigations.

We have analyzed MHD equations using symmetric variables which aided our diagrammatic analysis. Through coarse-graining we have generated a number of connected reducible diagrams, which were found to be marginal in the RG sense unless the scaling exponents are chosen in accordance with the fluctuation dissipation theorem. In fact we have shown that for MHD equations there are infinitely many such diagrams at the tree level. In practice this means that we cannot asses the behavior of such terms in the macroscopic limit, as they could either grow, decay or stay finite. This has a destructive effect on the RG analysis in general. In the case of a neutral fluid, we have looked at a particular regime, when velocity fluctuations are subject to thermal fluctuations. Such a system is in equilibrium and, as such, obeys equilibrium thermodynamics. By means of the fluctuation dissipation theorem which must hold, we have chosen the scaling exponents along the RG trajectory in such a way that the non-linear coupling is zero at a fixed point. As a result all non-linear effects are driven to zero. By considering various extensions of to Navier-Stokes, namely, MHD with kinetic and magnetic driving separately, we have limited our analysis to terms up to λ^2. The tree level terms are treated in a similar manner, by making a particular choice for the scaling exponents along the RG trajectory all non-linear effects can be driven to zero in the macroscopic limit. This justifies our choice of working with equation up to λ^2 only. Such choices for scaling exponents may seem artificial, and the question is whether or not physical systems resemble the linear behavior we predict perhaps can be best answered through numerical simulations. In any case, we are primarily concerned with the controlling tree level diagrams in our expansion and our treatment, at the very least, allows us to systematically avoid them.

Although, within our framework terms of order λ^3 are irrelevant, we believe that there is still a lot of worth studying. For one thing, we have shown that the loop correction to the three-point vertex introduces a new structure. The kinetic equation is renormalized while the induction equation is not. This is in partial agreement with [1], and disagrees with other calculations [2, 3]. A consistent methodology for self checks at this order would be the use of Ward-Takahashi identities, which provide relations between various three-point vertices. We have derived a particular example of that in the previous chapter, but in order to check the consistency of the third order results a more thorough study is needed. To our knowledge no such investigations has been done.

In any case, the problem of the reducible diagrams limits the scale of applications of our RG equations. Any alternative mechanism imposed on the couplings to suppress such terms would almost inevitably lead to a linear theory. The problem is ultimately linked to the fact that we work with reducible diagrams, which not only increase the complexity but, as in our case, represent problematic terms. Perhaps an alternative formulation, using irreducible representation, would solve the problem. There are no such calculations to date and this would most certainly be an interesting problem to look at. Alternatively, it would be interesting to apply the RG method which is often employed in QFT. The limitation of the method is that it does not predict phase-flows across the whole phase-space, but on the other hand it is a well established strict algorithm.

The application of RG techniques to turbulence remains an interesting field of research. Even though there are reports and papers which recover the Kolmogorov spectrum in fluid turbulence [4, 5] in one form or another, there is still no general consensus on the matter.

References

1. J.-D. Fournier, P.-L. Sulem, A. Pouquet, Infrared properties if forced magnetohydrodynamic turbulence. J. Phys. A: Math. Gen. **15**, 1393 (1982)
2. S.J. Camargo, H. Tasso, Renormalization group in magnetohydrodynamics turbulence, Phys. Fluids B, **4**(5), 1199–1212 (1992)
3. A. Berera, D. Hochberg, Asymptotic properties of turbulent magnetohydrodynamics, cond–mat/0103447 (2006)
4. D. Homeier, G. Mnster, Renormalization group analysis of turbulent hydrodynamics, arXiv/1012.0461 (2010)
5. C. de Dominicis, P.C. Martin, Energy spectra of certain randomly-stirred fluids. Phys. Rev. A **19**, 419 (1979)

Appendix A
Integral Identities

A.1 Standard Integrals

A.1.1 Multi-Dimensional Surface Integrals

Throughout our calculation of the loop integrals we come across a number of surface integrals which, due to their frequent occurrence, is useful to list. Consider a d-dimensional integral:

$$I = \int d^d\mathbf{x}. \tag{A.1.1}$$

We can split it into a radial part and a surface integral over a sphere:

$$\int d^d\mathbf{x} = \int_0^R r^{d-1}dr \int d\Omega, \tag{A.1.2}$$

where

$$\int d\Omega = \int_0^\pi \sin^{d-2}(\phi_1)\sin^{d-3}(\phi_2)\ldots\sin(\phi_{d-2})\,d\phi_1 d\phi_2 d\phi_{d-2}\int_0^{2\pi} d\phi_{d-1}. \tag{A.1.3}$$

Throughout, we will denote the surface integral as follows:

$$\int d\Omega = S_d, \tag{A.1.4}$$

E. Barkhudarov, *Renormalization Group Analysis of Equilibrium and Non-equilibrium Charged Systems*, Springer Theses,
DOI: 10.1007/978-3-319-06154-2,

where S_d is a surface of the d-dimensional unit sphere. Then, we can write the following two identities:

$$\int d\Omega \frac{q_i q_j}{q^2} = \delta_{ij} \frac{S_d}{d}, \tag{A.1.5}$$

$$\int d\Omega \frac{q_i q_j q_k q_m}{q^4} = \left(\delta_{ij}\delta_{km} + \delta_{ik}\delta_{jm} + \delta_{im}\delta_{jk}\right) \frac{S_d}{d\,(d+2)}. \tag{A.1.6}$$

A.1.2 Contour Integrals

The frequency integrals in the loop corrections are usually calculated by the residue method. Here we state the key integrals used through out the calculation:

$$\frac{1}{2\pi i} \int_{-\infty}^{\infty} dz \frac{1}{(z - ia)\left(z^2 + b^2\right)} = \frac{1}{2b\,(a+b)}, \tag{A.1.7}$$

$$\frac{1}{2\pi i} \int_{-\infty}^{\infty} dz \frac{1}{(z - ia)\left(z^2 + b^2\right)^2} = \frac{2b + a}{4b^3\,(b+a)^2}, \tag{A.1.8}$$

$$\frac{1}{2\pi} \int_{-\infty}^{\infty} dz \frac{1}{\left(z^2 + a^2\right)\left(z^2 + b^2\right)} = \frac{1}{2ab\,(a+b)}. \tag{A.1.9}$$

Throughout we assume that $a > 0$ and $b > 0$.

A.1.3 Heaviside Step Function and Symmetrization of Loop Integrals

Through out the calculation we come across the following integral:

$$I = \int_{>} \frac{d\mathbf{q}}{(2\pi)^d} q^{-y-4} \frac{q_i q_j q_k}{q^2}. \tag{A.1.10}$$

The domain of integration is restricted such that:

$$\Lambda e^{-r} < q < \Lambda, \tag{A.1.11}$$

and

$$\Lambda e^{-r} < |\mathbf{k} - \mathbf{q}| < \Lambda. \tag{A.1.12}$$

Using the second inequality, we can write the restriction in domain of integration to any order in k:

$$|\mathbf{k} - \mathbf{q}| \approx q - \cos\gamma k + \ldots \tag{A.1.13}$$

Thus to first order in k our inequalities are more restrictive:

$$\Lambda e^{-r} + \cos\gamma k < q < \Lambda, \qquad \cos\gamma > 0, \tag{A.1.14}$$

$$\Lambda e^{-r} < q < \Lambda + \cos\gamma k, \qquad \cos\gamma < 0. \tag{A.1.15}$$

The volume integral is then split according to the above inequalities;

$$\int_{>} d\mathbf{q} = \int_{\Lambda e^{-r}}^{\Lambda} dq \int d\Omega - \int_{\Lambda e^{-r}}^{\Lambda e^{-r}+\cos\gamma k} \int d\Omega_{+} - \int_{\Lambda+\cos\gamma k}^{\Lambda} \int d\Omega_{-}. \tag{A.1.16}$$

The contribution to the integral I comes from the split domain $d\Omega_{\pm}$. This is one way of doing the above integral. The result in the literature [1] we quote:

$$I = -\frac{1}{2d\,(d+2)} \frac{S_d}{(2\pi)^d} \Lambda^{-\varepsilon} \left(e^{\varepsilon r} - 1\right) \left(k_i \delta_{jk} + k_j \delta_{ik} + k_k \delta_{ij}\right). \tag{A.1.17}$$

An alternative way of calculating I is to insert the step functions explicitly to restrict the domain of integration [2]:

$$I = \int \frac{d\mathbf{q}}{(2\pi)^d} q^{-y-4} \frac{q_i q_j q_k}{q^2} \theta^{+}(q)\, \theta^{+}(|\mathbf{k} - \mathbf{q}|), \tag{A.1.18}$$

where we identify

$$\theta^{+}(q) = \theta\left(q - \Lambda e^{-r}\right) \theta\left(\Lambda - q\right) = \begin{cases} 1 & \Lambda e^{-r} < q < \Lambda, \\ 1/2 & \Lambda e^{-r} = q, \\ 0 & \text{otherwise}. \end{cases} \tag{A.1.19}$$

We then simply expand the step function in powers of k:

$$\theta^{+}(|\mathbf{k} - \mathbf{q}|) = \theta^{+}(q) + \frac{\mathbf{k}\cdot\mathbf{q}}{q} \left\{\delta\left(\Lambda - q\right) \theta\left(q - \Lambda e^{-r}\right) - \delta\left(q - \Lambda e^{-r}\right) \theta\left(\Lambda - q\right)\right\} + \ldots \tag{A.1.20}$$

We substitute the above into the volume integral:

$$I = \int q^{d-4-y} dq \int \frac{d\Omega}{(2\pi)^d} \frac{q_i q_j q_k}{q^3} \left[\theta^+ (q) \theta^+ (q) \right.$$
$$\left. + \theta^+ (q) \frac{\mathbf{k} \cdot \mathbf{q}}{q} \left\{ \delta (\Lambda - q) \theta \left(q - \Lambda e^{-r} \right) - \delta \left(q - \Lambda e^{-r} \right) \theta (\Lambda - q) \right\} \right]. \tag{A.1.21}$$

The first term vanishes. The second terms gives:

$$\begin{aligned} I =& p_m \int q^{d-4-y} dq \int \frac{d\Omega}{(2\pi)^d} \frac{q_i q_j q_k q_m}{q^4} \theta^+ (q) \delta (\Lambda - q) \theta \left(q - \Lambda e^{-r} \right) \\ &- p_m \int q^{d-4-y} dq \int \frac{d\Omega}{(2\pi)^d} \frac{q_i q_j q_k q_m}{q^4} \theta^+ (q) \delta \left(q - \Lambda e^{-r} \right) \theta (\Lambda - q) \\ =& p_m \Lambda^{d-4-y} \int \frac{d\Omega}{(2\pi)^d} \frac{q_i q_j q_k q_m}{q^4} \theta^+ (\Lambda) \theta \left(\Lambda - \Lambda e^{-r} \right) \\ &- p_m \left(\Lambda e^{-r} \right)^{d-4-y} \int \frac{d\Omega}{(2\pi)^d} \frac{q_i q_j q_k q_m}{q^4} \theta^+ \left(\Lambda e^{-r} \right) \theta \left(\Lambda - \Lambda e^{-r} \right) \\ =& p_m \frac{\Lambda^{d-4-y} - \left(\Lambda e^{-r} \right)^{d-4-y}}{2} \int \frac{d\Omega}{(2\pi)^d} \frac{q_i q_j q_k q_m}{q^4}. \end{aligned} \tag{A.1.22}$$

The last integral is a well known identity []:

$$\int \frac{d\Omega}{(2\pi)^d} \frac{q_i q_j q_k q_m}{q^4} = \frac{\left(\delta_{ij} \delta_{km} + \delta_{ik} \delta_{jm} + \delta_{im} \delta_{jk} \right)}{d (d+2)} \frac{S_d}{(2\pi)^d}. \tag{A.1.23}$$

Hence, we end up with the result:

$$I = \frac{\Lambda^{d-4-y} - \left(\Lambda e^{-r} \right)^{d-4-y}}{2d (d+2)} \frac{S_d}{(2\pi)^d} \left(p_k \delta_{ij} + p_j \delta_{jm} + p_i \delta_{jk} \right). \tag{A.1.24}$$

Using the notation in the literature:

$$I = -\frac{\left(e^{\varepsilon r} - 1 \right) \Lambda^{-\varepsilon}}{2d (d+2)} \frac{S_d}{(2\pi)^d} \left(p_k \delta_{ij} + p_j \delta_{jm} + p_i \delta_{jk} \right), \tag{A.1.25}$$

where $\varepsilon = 4 + y - d$. Although Camargo and Tasso did not manage to prove the above result for an arbitrary dimension [1] the above method clearly provides an advantage, since we work in arbitrary d from the outset.

Another advantage of keeping track of the step functions explicitly is when one tries to symmetrize the momentum distribution in the loop. Consider the second order graph, under a transformation:

$$\mathbf{q} \rightarrow \mathbf{q} + \frac{1}{2}\mathbf{k}, \tag{A.1.26}$$

then:

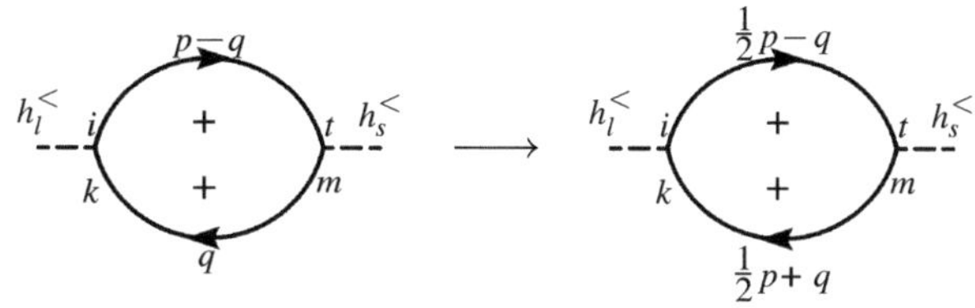

(A.1.27)

Under such a transformation of the integration variable the domain of integration is restricted to:

$$\theta^{+}(|\mathbf{q}|)\,\theta^{+}(|\mathbf{k}-\mathbf{q}|) \rightarrow \theta^{+}\left(\left|\mathbf{q}+\frac{1}{2}\mathbf{k}\right|\right)\theta^{+}\left(\left|\mathbf{q}-\frac{1}{2}\mathbf{k}\right|\right). \tag{A.1.28}$$

As a result there is no first order correction in $\mathbf{k}$ since upon a Taylor expansion it vanishes. Therefore there is no correction to integrals of the form:

$$\int \frac{d\Omega}{(2\pi)^d}\frac{q_i q_j q_k}{q^3}, \tag{A.1.29}$$

so they can be safely ignored. However, it is important to keep in mind that this follows only once the transformation $\mathbf{q} \rightarrow \mathbf{q} + \frac{1}{2}\mathbf{k}$ has been used, otherwise the above integral does contribute. This point has been more thoroughly explored in the context of Navier-Stokes RG [2] calculation but can be clearly an issue in MHD calculation as well.

The point on symmetrization of the loop integrals has been a source of numerous errors [3, 4] in the loop corrections of the Navier-Stokes equation. Since, the application of the Wilson-Kadanoff RG to MHD has provided inconsistent results in the past, we believe it is necessary to re-enforce the above strategy for clarity.

$$q = q' + \frac{1}{2}s \qquad \text{(A.1.20)}$$

then:

(A.1.21)

Under such a transformation of the integration variable the domain of integration is restricted to:

$$[illegible]$$

[illegible]

$$[illegible]$$

[illegible]

Appendix B
Second Order Loop Integrals

B.1 Noise Corrections

A typical graph, which contributes to noise corrections proportional to $A + B$ is computed from terms of the form:

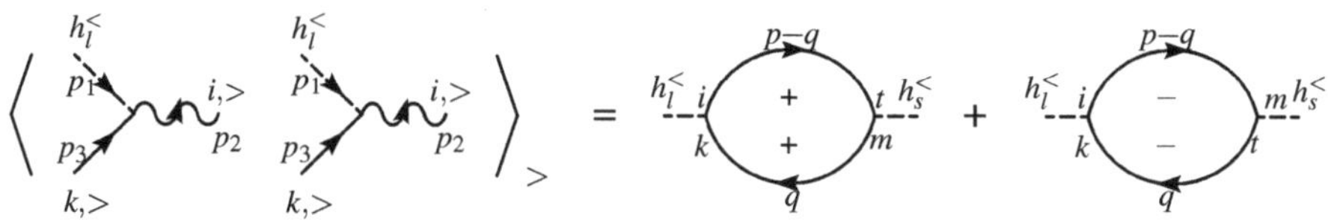

(B.1.1)

We have only included the connected diagrams. In the cumulant expansion all the disconnected diagrams are canceled. The expression on the RHS can be written mathematically as:

$$\lambda^2 \int\limits_{<} dp \mathrm{R}^<_{lik}(-\mathbf{p})\, \mathrm{R}^<_{stm}(\mathbf{p})\, h^<_l(p)\, h^<_s(-p)\, \boldsymbol{L}(p)\,, \tag{B.1.2}$$

where $L(p)$ is the loop correction. The above contribution is of order p^2, so we are only interested in the zero order contribution from the loop integral. After taking the limits $\mathbf{p}, \Omega \to 0$, we obtain:

$$\begin{aligned} L =& \int\limits_{>} \frac{dq}{(2\pi)^{d+1}} \left(G^+_{it}(-q)\, G^+_{km}(q) + G^-_{im}(-q)\, G^-_{kt}(q)\right) \\ =& 4 \int\limits_{>} \frac{d\mathbf{q}}{(2\pi)^d} \mathrm{P}_{it}(-\mathbf{q})\, \mathrm{P}_{km}(\mathbf{q}) \int\limits_{-\infty}^{\infty} \frac{d\omega}{2\pi} |\mathbf{q}|^{-2y} \left(\frac{A}{\omega^2 + \nu^2 q^4} + \frac{B}{\omega^2 + \mu^2 q^4}\right)^2 \\ &+ 4 \int\limits_{>} \frac{d\mathbf{q}}{(2\pi)^d} \mathrm{P}_{im}(-\mathbf{q})\, \mathrm{P}_{kt}(\mathbf{q}) \int\limits_{-\infty}^{\infty} \frac{d\omega}{2\pi} |\mathbf{q}|^{-2y} \left(\frac{A}{\omega^2 + \nu^2 q^4} - \frac{B}{\omega^2 + \mu^2 q^4}\right)^2 . \end{aligned} \tag{B.1.3}$$

E. Barkhudarov, *Renormalization Group Analysis of Equilibrium and Non-equilibrium Charged Systems*, Springer Theses,
DOI: 10.1007/978-3-319-06154-2, © Springer International Publishing Switzerland 2014

The integral over the frequency domain is performed by standard methods of contour integration. We can define the following contribution from the frequency integration:

$$\mathscr{A}_{\pm} = \left(\frac{A^2}{\nu^3} + \frac{B^2}{\mu^3} \pm \frac{2AB}{\nu\mu\,(\nu+\mu)} \right). \tag{B.1.4}$$

Thus, we have:

$$L = \mathscr{A}_{+} \int\limits_{>} \frac{d\mathbf{q}}{(2\pi)^d} |\mathbf{q}|^{-2y-6}\, \mathrm{P}_{it}\,(-\mathbf{q})\, \mathrm{P}_{km}\,(\mathbf{q}) + \mathscr{A}_{-} \int\limits_{>} \frac{d\mathbf{q}}{(2\pi)^d} |\mathbf{q}|^{-2y-6}\, \mathrm{P}_{im}\,(-\mathbf{q})\, \mathrm{P}_{kt}\,(\mathbf{q}). \tag{B.1.5}$$

Next, we make use of the definition of the projector operators to write them down explicitly:

$$\mathrm{P}_{it}\,(-\mathbf{q})\, \mathrm{P}_{km}\,(\mathbf{q}) = \delta_{it}\delta_{km} - \delta_{it}\frac{q_k q_m}{q^2} - \delta_{km}\frac{q_i q_t}{q^2} + \frac{q_i q_t q_k q_m}{q^4}, \tag{B.1.6}$$

$$\mathrm{P}_{im}\,(-\mathbf{q})\, \mathrm{P}_{kt}\,(\mathbf{q}) = \delta_{im}\delta_{kt} - \delta_{im}\frac{q_k q_t}{q^2} - \delta_{kt}\frac{q_i q_m}{q^2} + \frac{q_i q_t q_k q_m}{q^4}. \tag{B.1.7}$$

This leads to the expression:

$$L = \int\limits_{>} \frac{d\mathbf{q}}{(2\pi)^d} q^{-2y-6} \left[\mathscr{A}_{+} \left(\delta_{it}\delta_{km} - \delta_{it}\frac{q_k q_m}{q^2} - \delta_{km}\frac{q_i q_t}{q^2} + \frac{q_i q_t q_k q_m}{q^4} \right) + \mathscr{A}_{-} \left(\delta_{im}\delta_{kt} - \delta_{im}\frac{q_k q_t}{q^2} - \delta_{kt}\frac{q_i q_m}{q^2} + \frac{q_i q_t q_k q_m}{q^4} \right) \right]. \tag{B.1.8}$$

By making use of standard identities in Appendix **A.1**, the above expression can be integrated over fast Fourier modes to give:

$$L = \delta_{it}\delta_{km} \left[\mathscr{A}_{+} \frac{d^2-3}{d\,(d+2)} + \mathscr{A}_{-} \frac{1}{d\,(d+2)} \right] \frac{S_d}{(2\pi)^d} \frac{\Lambda^{\tilde{\varepsilon}}}{\tilde{\varepsilon}} \left(1 - b^{-\tilde{\varepsilon}} \right), \tag{B.1.9}$$

$$\tilde{\varepsilon} = d - 2y - 6. \tag{B.1.10}$$

By substituting this result to the original expression we obtain our final contribution:

$$\rightarrow -\frac{1}{2}\lambda^2 L \int\limits_{<} dp p^2 \mathrm{P}_{ls}\,(\mathbf{p})\, h_l^{<}\,(p)\, h_s^{<}\,(-p). \tag{B.1.11}$$

The factor of a half comes as a result of the cumulant expansion. Since it is a second order term, it has such a prefactor. In the above form is how this term appears as a new contribution to the action.

Further, we must to calculate cross-correlations between conjugate fields. First, corrections of such form appear at the second order in λ. The diagrammatic representation of a typical term is:

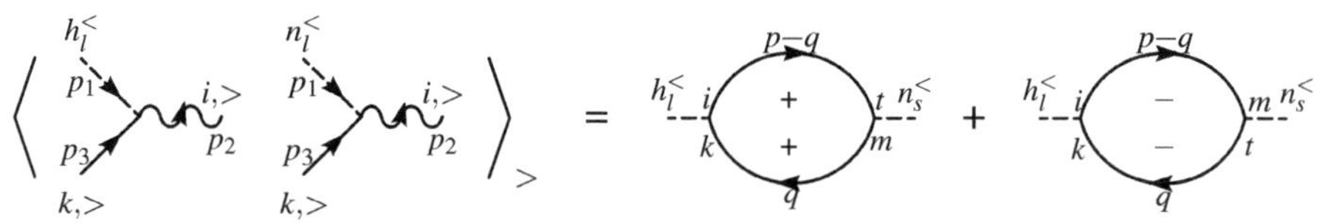

It is clear that the loop contribution is identical to the one we have already calculated with the only difference that we have interchanged $\mathscr{A}_+ \leftrightarrow \mathscr{A}_-$. Thus, we can immediately write the contribution as:

$$- \lambda^2 \tilde{L} \int_< dpp^2 \mathrm{P}_{ls} (\mathbf{p}) \, h_l^< (p) \, n_s^< (-p) , \tag{B.1.12}$$

where

$$\tilde{L} = \delta_{it} \delta_{km} \left[\mathscr{A}_- \frac{d^2 - 3}{d (d+2)} + \mathscr{A}_+ \frac{1}{d (d+2)} \right] \frac{S_d}{(2\pi)^d} \frac{\Lambda^{\tilde{\varepsilon}}}{\tilde{\varepsilon}} \left(1 - b^{-\tilde{\varepsilon}} \right) \tag{B.1.13}$$

Note, there is an additional factor of 2, which cancels the half. This is a consequence of the fact that at the second order in λ all of the non-identical terms pick up a factor of 2 from the quadratic expansion.

The terms which have been generated as a result of coarse graining are proportional to p^2, where, the original part of the action which contains the noise matrix is proportional to p^{-y}. Unless we set $y = -2$, we cannot write the above contribution as a correction to the noise amplitude in closed form. We can interpret the appearance of this new term within the RG framework naturally as a new interaction which appear upon coarse-graining. We have started with a theory which had this particular interaction switched off and then generated it upon coarse-graining. Accordingly, if we were to include this interaction from the outset we would be able to write down a closed form expression for the correction to the noise amplitude. At the same time we could identify a boundary condition for this new term such that it is zero before we coarse-grain the system and thereby formally retain the problem as it is originally formulated.

Hence, we proceed by a redefinition of the noise amplitude:

$$Ak^{-y} \rightarrow A(k) = \left(A_0 + k^{2+y} A_2 \right) k^{-y} , \tag{B.1.14}$$

$$Bk^{-y} \to B(k) = \left(B_0 + k^{2+y} B_2\right) k^{-y}. \tag{B.1.15}$$

Upon this substitution the loop integral is:

$$L = \int\limits_{>} \frac{d\mathbf{q}}{(2\pi)^d} \mathscr{A}_+ (|\mathbf{q}|) \, |\mathbf{q}|^{-2y-6} \, \mathrm{P}_{it}(-\mathbf{q}) \, \mathrm{P}_{km}(\mathbf{q})$$
$$+ \int\limits_{>} \frac{d\mathbf{q}}{(2\pi)^d} \mathscr{A}_- (|\mathbf{q}|) \, |\mathbf{q}|^{-2y-6} \, \mathrm{P}_{im}(-\mathbf{q}) \, \mathrm{P}_{kt}(\mathbf{q}) \, . \tag{B.1.16}$$

Such integral would not change any angular integrations. In fact, upon a Taylor expansion of the integral in radial direction around Λ we obtain:

$$L = \delta_{it}\delta_{km} \left[\mathscr{A}_+ (\Lambda) \frac{d^2 - 3}{d(d+2)} + \mathscr{A}_- (\Lambda) \frac{1}{d(d+2)} \right] \frac{S_d}{(2\pi)^d} \Lambda^{\tilde{\varepsilon}} \delta\ell, \tag{B.1.17}$$

and, for the cross correlation, we can write:

$$\tilde{L} = \delta_{it}\delta_{km} \left[\mathscr{A}_- (\Lambda) \frac{d^2 - 3}{d(d+2)} + \mathscr{A}_+ (\Lambda) \frac{1}{d(d+2)} \right] \frac{S_d}{(2\pi)^d} \Lambda^{\tilde{\varepsilon}} \delta\ell, \tag{B.1.18}$$

where we have defined:

$$\mathscr{A}_\pm (\Lambda) = \frac{\left(A_0 + \Lambda^{2+y} A_2\right)^2}{\nu^3} + \frac{\left(B_0 + \Lambda^{2+y} B_2\right)^2}{\mu^3}$$
$$\pm \frac{2\left(A_0 + \Lambda^{2+y} A_2\right)\left(B_0 + \Lambda^{2+y} B_2\right)}{\nu\mu(\nu + \mu)}. \tag{B.1.19}$$

We have made use of the following definition:

$$b = e^{\delta\ell} \approx 1 + \delta\ell + \dots . \tag{B.1.20}$$

Having re-defined the noise we have generated a closed-form corrections to the noise amplitudes.

B.2 Corrections to γ_+

Corrections to the coefficient γ_+ is given by the following contributions:

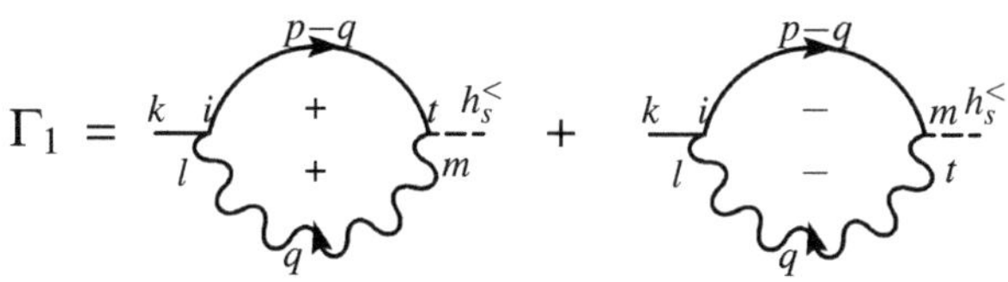

(B.2.1)

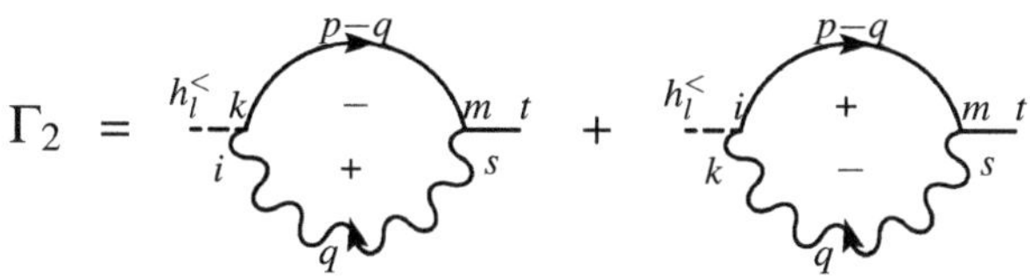

(B.2.2)

The first contribution reads:

$$\Gamma_1 = -i \int dp \mathrm{R}_{stm}\left(-\mathbf{p}\right) h_s\left(p\right) P_k\left(-p\right) \gamma_1\left(\mathbf{p}, \Omega \to 0\right), \tag{B.2.3}$$

where the loop correction is:

$$\begin{aligned}
\gamma_1 &= \int\limits_{>} dq \mathrm{R}_{lik}\left(-\mathbf{q}\right)\left[G^{+}_{it}\left(p-q\right) C^{+}_{lm}\left(q\right) + G^{-}_{im}\left(p-q\right) C^{-}_{lt}\left(q\right)\right] \\
&= \int\limits_{>} dq \mathrm{R}_{lik}\left(-\mathbf{q}\right) \mathrm{P}_{im}\left(p-q\right) g\left(p-q\right) \delta_{lt}\left(\frac{1}{i\omega + \nu q^2} - \frac{1}{i\omega + \mu q^2}\right) \\
&\quad \times \left(\frac{A\left(p-q\right)}{\omega^2 + \nu^2\left(p-q\right)^4} - \frac{B\left(p-q\right)}{\omega^2 + \mu^2\left(p-q\right)^4}\right) \\
&\quad + \int\limits_{>} dq \mathrm{R}_{lik}\left(-\mathbf{q}\right) \mathrm{P}_{it}\left(p-q\right) g\left(p-q\right) \delta_{lm}\left(\frac{1}{i\omega + \nu q^2} + \frac{1}{i\omega + \mu q^2}\right) \\
&\quad \times \left(\frac{A\left(p-q\right)}{\omega^2 + \nu^2\left(p-q\right)^4} + \frac{B\left(p-q\right)}{\omega^2 + \mu^2\left(p-q\right)^4}\right).
\end{aligned} \tag{B.2.4}$$

The correction to the coefficient is of the order p^2, thus we seek to expand the above expression in orders of $\mathbf{p}$ to extract the leading order contribution from the coarse-graining. We come across a frequency integral which is evaluated using the method of residues:

$$\begin{aligned}
&\int\limits_{-\infty}^{\infty} \frac{d\omega}{2\pi} \frac{1}{\left(i\omega + \mu q^2\right)\left(\omega^2 + \nu^2\left(p-q\right)^4\right)} \\
&\qquad = \frac{1}{q^4}\left(\frac{1}{2\nu\left(\nu+\mu\right)} + \frac{2\nu+\mu}{\nu\left(\nu+\mu\right)^2}\left(\frac{\mathbf{p}\cdot\mathbf{q}}{q^2}\right)\right) + \dots .
\end{aligned} \tag{B.2.5}$$

After algebraic manipulations the contribution from γ_1 can be compactly written using the following notation:

$$\mathscr{B}_{\pm} = \frac{1}{4}\left(\frac{A}{\nu^2} + \frac{B}{\mu^2}\right) \pm \frac{1}{2(\nu+\mu)}\left(\frac{A}{\nu} + \frac{B}{\mu}\right), \tag{B.2.6}$$

$$\mathscr{C}_{\pm} = A\left(\frac{3}{4\nu^2} \pm \frac{2\nu+\mu}{\nu(\nu+\mu)^2}\right) + B\left(\frac{3}{4\mu^2} \pm \frac{2\mu+\nu}{\mu(\nu+\mu)^2}\right). \tag{B.2.7}$$

Then:

$$\begin{aligned}\gamma_1 = \int_{>} \frac{d\mathbf{q}}{(2\pi)^d} \Bigg[&\mathrm{R}_{lik}(-\mathbf{q})\,\mathrm{P}_{im}(p-q)\, g(p-q)\, \delta_{lt} q^{-4}\left(\mathscr{B}_- + \mathscr{C}_-\left(\frac{\mathbf{p}\cdot\mathbf{q}}{q^2}\right)\right) \\ &+ \mathrm{R}_{lik}(-\mathbf{q})\,\mathrm{P}_{it}(p-q)\, g(p-q)\, \delta_{lm} q^{-4}\left(\mathscr{B}_+ + \mathscr{C}_+\left(\frac{\mathbf{p}\cdot\mathbf{q}}{q^2}\right)\right)\Bigg]. \end{aligned} \tag{B.2.8}$$

Note that the coefficients $\mathscr{B}_{\pm}$ and $\mathscr{C}_{\pm}$ also carry momentum dependence through $A(p-q)$ and $B(p-q)$. We have omitted this dependence for compactness. We proceed by expanding the projector operator and the function g in powers of $\mathbf{p}$. The projector operator can be written to first order as:

$$\mathrm{P}_{im}(p-q) = \mathrm{P}_{im}(q) + \mathscr{O}_{mi}(p,q) + \ldots \tag{B.2.9}$$

The second term is of order p, explicitly it has the form:

$$\mathscr{O}_{mi}(p,q) = \frac{1}{q^2}\left(p_i q_m + q_i p_m - q_i q_m 2\frac{\mathbf{p}\cdot\mathbf{q}}{q^2}\right), \tag{B.2.10}$$

and

$$g(|\mathbf{p}-\mathbf{q}|) = q^{-y}\left(1 + y\frac{\mathbf{p}\cdot\mathbf{q}}{q^2} + \ldots\right). \tag{B.2.11}$$

It follows that:

$$\mathrm{R}_{tik}(-\mathbf{q})\,\mathrm{P}_{im}(p-q)\, g(p-q) = q^{-y}\left(-p_j \frac{q_t q_k q_m q_j}{q^4}\right) + \cdots \tag{B.2.12}$$

The above expression follows from properties of the projector operator and the fact that we have solenoidal fields. A similar analysis leads to an analogous contribution in the second term:

$$\begin{aligned}&\mathrm{R}_{mik}(-\mathbf{q})\,\mathrm{P}_{it}(p-q)\, g(p-q) \\ &= q^{-y}\left(p_t \frac{q_m q_k}{q^2} - \delta_{mk} p_t - p_j \frac{q_m q_k q_t q_j}{q^4} + \delta_{mk} p_j \frac{q_t q_j}{q^2}\right). \end{aligned} \tag{B.2.13}$$

An important fact about the above expansions is that they are of order p and hence we do not need to go beyond the zero-order expansion of the other terms. The contribution from the graphs then reads:

$$\gamma_1 = \int_{>} \frac{d\mathbf{q}}{(2\pi)^d} q^{-4-y} \left[\left(p_t \frac{q_m q_k}{q^2} - \delta_{mk} p_t - p_j \frac{q_m q_k q_t q_j}{q^4} + \delta_{mk} p_j \frac{q_t q_j}{q^2} \right) \right.$$

$$\left. \mathscr{B}_+ (q) - p_j \frac{q_t q_k q_m q_j}{q^4} \mathscr{B}_- (q) \right]. \tag{B.2.14}$$

The surface integrals are evaluated by means of identities we have identified earlier. After some algebra we obtain:

$$\gamma_1 = \delta_{mk} p_t \left[\mathscr{B}_+ (\Lambda) \frac{3 - d^2}{d(d+2)} - \mathscr{B}_- (\Lambda) \frac{1}{d(d+2)} \right] \frac{S_d}{(2\pi)^d} \Lambda^{\varepsilon} \delta \ell, \tag{B.2.15}$$

$$\varepsilon = d - 4 - y. \tag{B.2.16}$$

The second correction to the coefficient γ_+ is:

$$\Gamma_2 = -i \int dp \mathrm{R}_{stm} (-\mathbf{p}) h_s (p) P_i (-p) \gamma_2 (\mathbf{p}, \Omega \to 0), \tag{B.2.17}$$

where the loop correction is:

$$\gamma_2 = \int_{>} dq \mathrm{R}_{lik} (-\mathbf{q}) \mathrm{P}_{mk} (p - q) g (p - q) \delta_{lt} \left(\frac{1}{i\omega + \nu q^2} + \frac{1}{i\omega + \mu q^2} \right)$$

$$\times \left(\frac{A (p - q)}{\omega^2 + \nu^2 (p - q)^4} - \frac{B (p - q)}{\omega^2 + \mu^2 (p - q)^4} \right)$$

$$+ \int_{>} dq \mathrm{R}_{lik} (-\mathbf{q}) \mathrm{P}_{tk} (p - q) g (p - q) \delta_{lm} \left(\frac{1}{i\omega + \nu q^2} - \frac{1}{i\omega + \mu q^2} \right)$$

$$\times \left(\frac{A (p - q)}{\omega^2 + \nu^2 (p - q)^4} + \frac{B (p - q)}{\omega^2 + \mu^2 (p - q)^4} \right). \tag{B.2.18}$$

Once again we introduce two definitions to write the frequency integrals in a compact manner:

$$\tilde{\mathscr{B}}_{\pm} = \frac{1}{4} \left(\frac{A}{\nu^2} - \frac{B}{\mu^2} \right) \pm \frac{1}{2(\nu + \mu)} \left(\frac{A}{\nu} - \frac{B}{\mu} \right), \tag{B.2.19}$$

$$\tilde{\mathscr{C}}_{\pm} = A \left(\frac{3}{4\nu^2} \pm \frac{2\nu + \mu}{\nu (\nu + \mu)^2} \right) - B \left(\frac{3}{4\mu^2} \pm \frac{2\mu + \nu}{\mu (\nu + \mu)^2} \right). \tag{B.2.20}$$

It follows that the loop contribution can be written as:

$$\gamma_2 = \int\limits_{>} \frac{d\mathbf{q}}{(2\pi)^d} \Bigg[\mathrm{R}_{lik}(-\mathbf{q}) \mathrm{P}_{km}(p-q) g(p-q) \delta_{lt} q^{-4} \left(\tilde{\mathscr{B}}_+ + \tilde{\mathscr{C}}_+ \left(\frac{\mathbf{p}\cdot\mathbf{q}}{q^2} \right) \right)$$

$$+ \mathrm{R}_{lik}(-\mathbf{q}) \mathrm{P}_{tk}(p-q) g(p-q) \delta_{lm} q^{-4} \left(\tilde{\mathscr{B}}_- + \tilde{\mathscr{C}}_- \left(\frac{\mathbf{p}\cdot\mathbf{q}}{q^2} \right) \right) \Bigg]. \quad \text{(B.2.21)}$$

We proceed by expanding the projectors:

$$\mathrm{R}_{lik}(-\mathbf{q}) \mathrm{P}_{km}(p-q) g(p-q) \delta_{lt}$$
$$= -q^{-y} \left[p_t \frac{q_i q_m}{q^2} - \frac{q_m q_t q_i}{q^2} - (1+y) p_k \frac{q_i q_k q_t q_m}{q^4} \right], \quad \text{(B.2.22)}$$

and

$$\mathrm{R}_{lik}(-\mathbf{q}) \mathrm{P}_{tk}(p-q) g(p-q) \delta_{lm} = q^{-y} \left[\frac{q_m q_t q_i}{q^2} + (1+y) p_k \frac{q_i q_k q_t q_m}{q^4} \right]. \quad \text{(B.2.23)}$$

We substitute these results into the momentum integral:

$$\gamma_2 = -\int\limits_{>} \frac{d\mathbf{q}}{(2\pi)^d} q^{-y-4} \left[\left(p_t \frac{q_i q_m}{q^2} - (1+y) p_k \frac{q_i q_k q_t q_m}{q^4} \right) \tilde{\mathscr{B}}_+ - \tilde{\mathscr{C}}_+ p_k \frac{q_i q_k q_t q_m}{q^4} \right]$$

$$+ \int\limits_{>} \frac{d\mathbf{q}}{(2\pi)^d} q^{-y-4} \left[(1+y) p_k \frac{q_i q_k q_t q_m}{q^4} \tilde{\mathscr{B}}_- + \tilde{\mathscr{C}}_- p_k \frac{q_i q_k q_t q_m}{q^4} \right]$$

$$+ \int\limits_{>} \frac{d\mathbf{q}}{(2\pi)^d} q^{-y-4} \frac{q_m q_t q_i}{q^2} \left[\tilde{\mathscr{B}}_- + \tilde{\mathscr{B}}_+ \right]. \quad \text{(B.2.24)}$$

We have written the last term separately because it requires a little bit more attention. We have to do a Taylor expansion around the momentum shells as well as expand the coefficients $\tilde{\mathscr{B}}_-$ and $\tilde{\mathscr{B}}_-$to first order on p. We have gone through how to evaluate the above integral upon momentum shell expansion in Appendix **A**. So here we solely focus on expansion of the coefficients. First, note:

$$\tilde{\mathscr{B}}_- + \tilde{\mathscr{B}}_+ = \frac{1}{2} \left(\frac{A(p-q)}{\nu^2} - \frac{B(p-q)}{\mu^2} \right) \quad \text{(B.2.25)}$$

As we Taylor expand this expression to first order in p, we have:

$$\tilde{\mathscr{B}}_- + \tilde{\mathscr{B}}_+ = \frac{1}{2} \left(\frac{A(q)}{\nu^2} - \frac{B(q)}{\mu^2} \right) - (2+y) \left(\frac{A_2 q^{2+y}}{2\nu^2} - \frac{B_2 q^{2+y}}{2\mu^2} \right) \frac{\mathbf{p}\cdot\mathbf{q}}{q^2} \quad \text{(B.2.26)}$$

Now we substitute that back into our expression for the loop integral:

$$\gamma_2 = -\int\limits_{>} \frac{d\mathbf{q}}{(2\pi)^d} q^{-y-4} \left[\left(p_t \frac{q_i q_m}{q^2} - \frac{q_m q_t q_i}{q^2} - (1+y)\, p_k \frac{q_i q_k q_t q_m}{q^4} \right) \right.$$
$$\left. \tilde{\mathscr{B}}_+ - \tilde{\mathscr{C}}_+ p_k \frac{q_i q_k q_t q_m}{q^4} \right]$$
$$+ \int\limits_{>} \frac{d\mathbf{q}}{(2\pi)^d} q^{-y-4} \left[\left(\frac{q_m q_t q_i}{q^2} + (1+y)\, p_k \frac{q_i q_k q_t q_m}{q^4} \right) \right.$$
$$\left. \tilde{\mathscr{B}}_- + \tilde{\mathscr{C}}_- p_k \frac{q_i q_k q_t q_m}{q^4} \right]$$
$$- \frac{(2+y)}{2} \int\limits_{>} \frac{d\mathbf{q}}{(2\pi)^d} q^{-2} \left(\frac{A_2}{\nu^2} - \frac{B_2}{\mu^2} \right) p_k \frac{q_m q_t q_i q_k}{q^4}. \tag{B.2.27}$$

The coefficients $\tilde{\mathscr{B}}_\pm$ and $\tilde{\mathscr{C}}_\pm$ in the first two lines are evaluated to the zeroth order in p and the first order contribution is captured by the last line. Upon integration of the surface integrals and a Taylor expansion of the integral in radial direction the final results read:

$$\gamma_2 = p_t \delta_{im} \frac{S_d}{(2\pi)^d} \delta\ell \Lambda^\varepsilon \left[-\frac{(d+6-y)}{2d\,(d+2)} \tilde{\mathscr{B}}_+^0 + \frac{d-2+y}{2d\,(d+2)} \tilde{\mathscr{B}}_-^0 + \frac{1}{d\,(d+2)} \left(\tilde{\mathscr{C}}_+^0 + \tilde{\mathscr{C}}_-^0 \right) \right]$$
$$+ p_t \delta_{im} \frac{S_d}{(2\pi)^d} \delta\ell \Lambda^{d-2} \left[-\frac{(d+4-2y)}{2d\,(d+2)} \tilde{\mathscr{B}}_+^2(\Lambda) + \frac{d+2y}{2d\,(d+2)} \tilde{\mathscr{B}}_-^2(\Lambda) \right.$$
$$\left. + \frac{1}{d\,(d+2)} \left(\tilde{\mathscr{C}}_+^2(\Lambda) + \tilde{\mathscr{C}}_-^2(\Lambda) \right) - \frac{(2+y)}{d\,(d+2)} \left(\tilde{\mathscr{B}}_+^2 + \tilde{\mathscr{B}}_-^2 \right) \right]. \tag{B.2.28}$$

We have calculated all relevant corrections to γ_+. Note that all our graphs contributed to corrections which are proportional to the h and P fields. An Analogous calculation would follow for graphs whose external legs are n and Q. From the diagrammatic construction it is clear that the corrections are identical.

B.3 Corrections to γ_-

The diagrammatic representation of the terms which contribute to corrections of γ_- is:

$\Phi_1 =$ 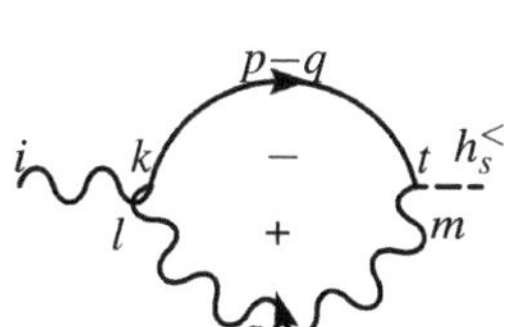 $+$ 

(B.3.1)

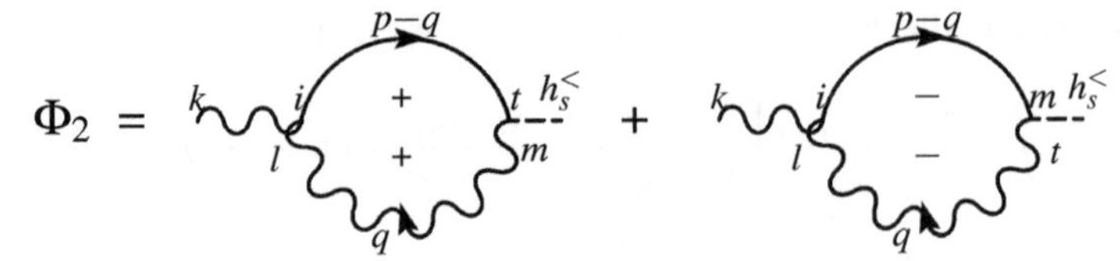

(B.3.2)

The first two diagrams when expressed mathematically are:

$$\Phi_1 = -i \int dp \mathrm{R}_{stm}(-\mathbf{p})\, h_s(p)\, Q_i(-p)\, \phi_1(\mathbf{p}, \Omega \to 0), \tag{B.3.3}$$

where the loop correction is:

$$\begin{aligned}\phi_1 = & \int\limits_{>} dq \mathrm{R}_{lik}(-\mathbf{q})\, \mathrm{P}_{mk}(p-q)\, g(p-q)\, \delta_{lt} \left(\frac{1}{i\omega + \nu q^2} - \frac{1}{i\omega + \mu q^2} \right) \\ & \times \left(\frac{A(p-q)}{\omega^2 + \nu^2 (p-q)^4} + \frac{B(p-q)}{\omega^2 + \mu^2 (p-q)^4} \right) \\ & + \int\limits_{>} dq \mathrm{R}_{lik}(-\mathbf{q})\, \mathrm{P}_{tk}(p-q)\, g(p-q)\, \delta_{lm} \left(\frac{1}{i\omega + \nu q^2} + \frac{1}{i\omega + \mu q^2} \right) \\ & \times \left(\frac{A(p-q)}{\omega^2 + \nu^2 (p-q)^4} - \frac{B(p-q)}{\omega^2 + \mu^2 (p-q)^4} \right). \end{aligned} \tag{B.3.4}$$

Integration over the frequency integral is the analogous to the calculation for γ_2 coefficient. The result reads:

$$\begin{aligned}\phi_1 = & \int\limits_{>} \frac{d\mathbf{q}}{(2\pi)^d} \left[\mathrm{R}_{lik}(-\mathbf{q})\, \mathrm{P}_{km}(p-q)\, g(p-q)\, \delta_{lt} q^{-4} \left(\tilde{\mathscr{B}}_- + \tilde{\mathscr{C}}_- \left(\frac{\mathbf{p}\cdot\mathbf{q}}{q^2} \right) \right) \right. \\ & \left. + \mathrm{R}_{lik}(-\mathbf{q})\, \mathrm{P}_{tk}(p-q)\, g(p-q)\, \delta_{lm} q^{-4} \left(\tilde{\mathscr{B}}_+ + \tilde{\mathscr{C}}_+ \left(\frac{\mathbf{p}\cdot\mathbf{q}}{q^2} \right) \right) \right]. \end{aligned} \tag{B.3.5}$$

Note that the projector operators have the same index dependence as in the Γ_2 contribution. The only difference is that we have swapped frequency integrals $\mathscr{B}_+ \leftrightarrow \mathscr{B}_-$ and $\mathscr{C}_+ \leftrightarrow \mathscr{C}_-$. Thus we can write:

$$\begin{aligned}\phi_1 = & -\int\limits_{>} \frac{d\mathbf{q}}{(2\pi)^d} q^{-y-4} \left[\left(p_t \frac{q_i q_m}{q^2} - \frac{q_m q_t q_i}{q^2} - (1+y)\, p_k \frac{q_i q_k q_t q_m}{q^4} \right) \tilde{\mathscr{B}}_- - \tilde{\mathscr{C}}_- p_k \frac{q_i q_k q_t q_m}{q^4} \right] \\ & + \int\limits_{>} \frac{d\mathbf{q}}{(2\pi)^d} q^{-y-4} \left[\left(\frac{q_m q_t q_i}{q^2} + (1+y)\, p_k \frac{q_i q_k q_t q_m}{q^4} \right) \tilde{\mathscr{B}}_+ + \tilde{\mathscr{C}}_+ p_k \frac{q_i q_k q_t q_m}{q^4} \right] \\ & - \frac{(2+y)}{2} \int\limits_{>} \frac{d\mathbf{q}}{(2\pi)^d} q^{-2} \left(\frac{A_2}{\nu^2} - \frac{B_2}{\mu^2} \right) p_k \frac{q_m q_t q_i q_k}{q^4}. \end{aligned} \tag{B.3.6}$$

We can immediately state the final result:

$$
\begin{aligned}
\phi_1 =& p_t \delta_{im} \frac{S_d}{(2\pi)^d} \delta\ell \Lambda^{\varepsilon} \left[-\frac{(d+6-y)}{2d\,(d+2)} \tilde{\mathscr{B}}_-^0 + \frac{d-2+y}{2d\,(d+2)} \tilde{\mathscr{B}}_+^0 + \frac{1}{d\,(d+2)} \left(\tilde{\mathscr{C}}_+^0 + \tilde{\mathscr{C}}_-^0 \right) \right] \\
&+ p_t \delta_{im} \frac{S_d}{(2\pi)^d} \delta\ell \Lambda^{d-2} \left[-\frac{(d+4-2y)}{2d\,(d+2)} \tilde{\mathscr{B}}_+^2(\Lambda) + \frac{d+2y}{2d\,(d+2)} \tilde{\mathscr{B}}_-^2(\Lambda) \right. \\
&\left. + \frac{1}{d\,(d+2)} \left(\tilde{\mathscr{C}}_+^2(\Lambda) + \tilde{\mathscr{C}}_-^2(\Lambda) \right) - \frac{(2+y)}{d\,(d+2)} \left(\tilde{\mathscr{B}}_+^2 + \tilde{\mathscr{B}}_-^2 \right) \right].
\end{aligned}
\tag{B.3.7}
$$

The second contribution is Φ_2 has the algebraic form:

$$
\Phi_2 = -i \int dp \mathrm{R}_{stm}(-\mathbf{p})\, h_s(p)\, Q_k(-p)\, \phi_2(\mathbf{p}, \Omega \to 0),
\tag{B.3.8}
$$

where the loop correction is:

$$
\begin{aligned}
\phi_2 =& \int_{>} dq \mathrm{R}_{lik}(-\mathbf{q})\, \mathrm{P}_{im}(p-q)\, g(p-q)\, \delta_{lt} \left(\frac{1}{i\omega + \nu q^2} + \frac{1}{i\omega + \mu q^2} \right) \\
&\times \left(\frac{A(p-q)}{\omega^2 + \nu^2 (p-q)^4} + \frac{B(p-q)}{\omega^2 + \mu^2 (p-q)^4} \right) \\
&+ \int_{>} dq \mathrm{R}_{lik}(-\mathbf{q})\, \mathrm{P}_{it}(p-q)\, g(p-q)\, \delta_{lm} \left(\frac{1}{i\omega + \nu q^2} - \frac{1}{i\omega + \mu q^2} \right) \\
&\times \left(\frac{A(p-q)}{\omega^2 + \nu^2 (p-q)^4} - \frac{B(p-q)}{\omega^2 + \mu^2 (p-q)^4} \right).
\end{aligned}
\tag{B.3.9}
$$

After performing the frequency integral we obtain:

$$
\begin{aligned}
\phi_2 =& \int_{>} \frac{d\mathbf{q}}{(2\pi)^d} \left[\mathrm{R}_{lik}(-\mathbf{q})\, \mathrm{P}_{im}(p-q)\, g(p-q)\, \delta_{lt} q^{-4} \left(\mathscr{B}_+ + \mathscr{C}_+ \left(\frac{\mathbf{p}\cdot\mathbf{q}}{q^2} \right) \right) \right. \\
&\left. + \mathrm{R}_{lik}(-\mathbf{q})\, \mathrm{P}_{it}(p-q)\, g(p-q)\, \delta_{lm} q^{-4} \left(\mathscr{B}_- + \mathscr{C}_- \left(\frac{\mathbf{p}\cdot\mathbf{q}}{q^2} \right) \right) \right].
\end{aligned}
\tag{B.3.10}
$$

This is exactly the same expression we have had for γ_1, only we have swapped $\mathscr{B}_+ \leftrightarrow \mathscr{B}_-$ and $\mathscr{C}_+ \leftrightarrow \mathscr{C}_-$. Hence, we obtain

$$
\phi_2 = \delta_{mk} p_t \left[\mathscr{B}_-(\Lambda) \frac{3-d^2}{d\,(d+2)} - \mathscr{B}_+(\Lambda) \frac{1}{d\,(d+2)} \right] \frac{S_d}{(2\pi)^d} \Lambda^{\varepsilon} \delta\ell.
\tag{B.3.11}
$$

Appendix C
Third Order Loop Integrals

C.1 Corrections to λ

We begin by considering all graphs which form a correction to the vertex:

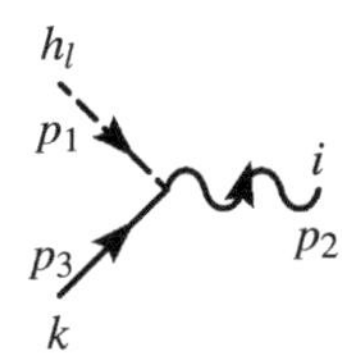

To account for all possible graphs we need to form combinations of one-loop graphs from the following five terms:

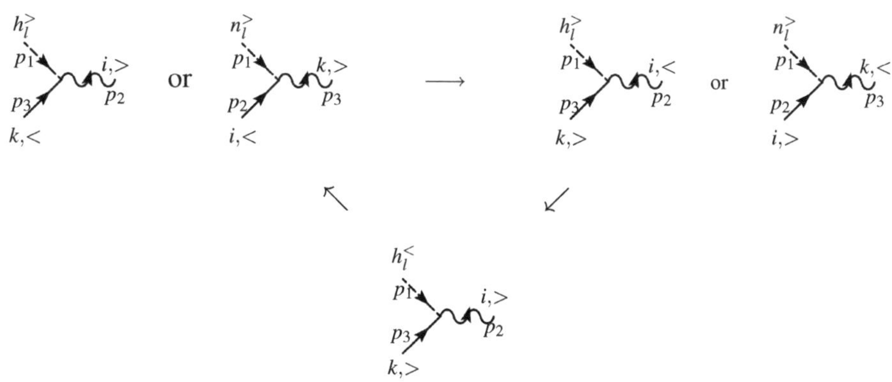

We choose either of the two terms which come in pair and combine it with either of the terms in the second pair. The third term is fixed, as there is only one graph of its kind. For any given triplet of terms we choose to combine we generate eight graphs. This number arises simply from counting all possible ways we can contract any given triplet.

E. Barkhudarov, *Renormalization Group Analysis of Equilibrium and Non-equilibrium Charged Systems*, Springer Theses,
DOI: 10.1007/978-3-319-06154-2,

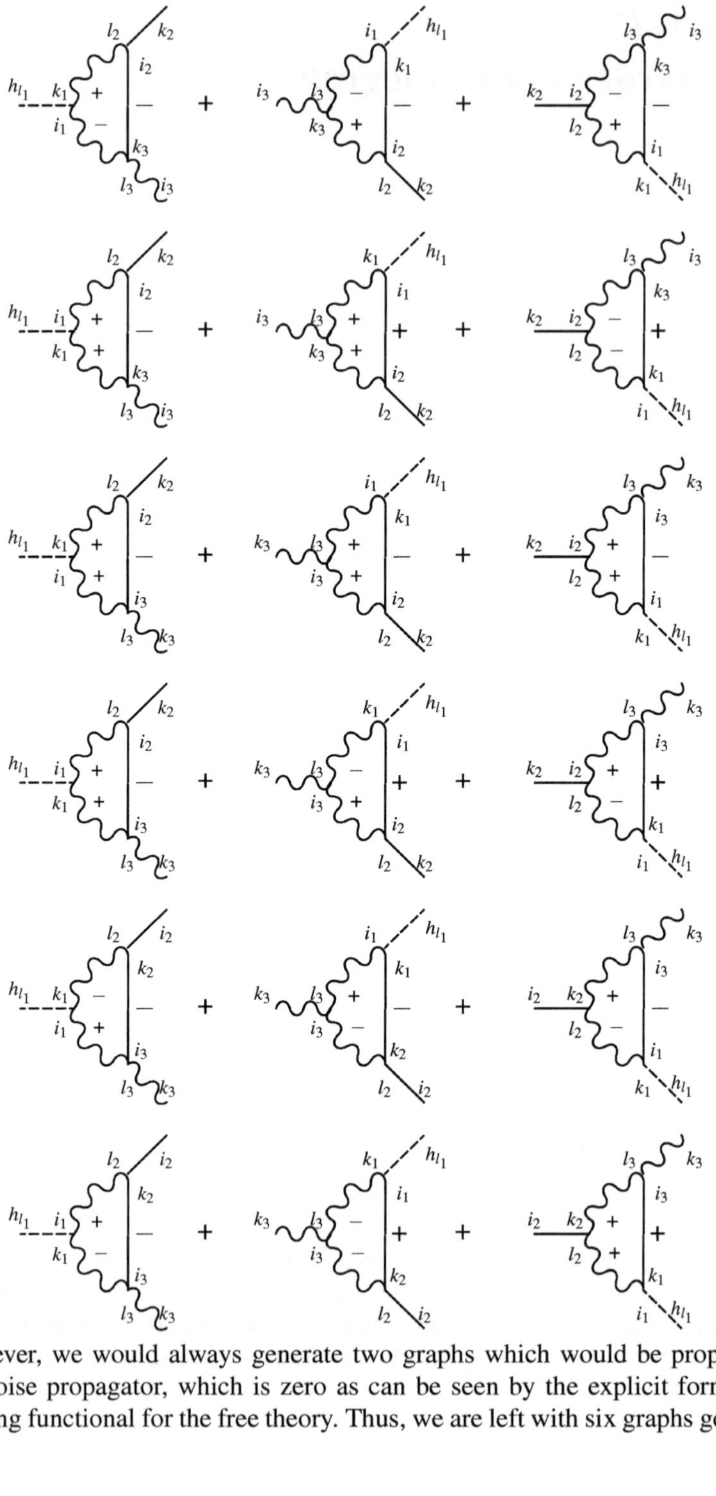

However, we would always generate two graphs which would be proportional to the noise propagator, which is zero as can be seen by the explicit form of the generating functional for the free theory. Thus, we are left with six graphs generated by each triplet. All together we have twenty four graphs. Consider the first eighteen graphs (see previous page). These diagrams all have the same property, namely, they vanish purely because of the way the indicies are distributed in the loop integral. This is only true in the limit of when the external momentum goes to zero. The presence of the dashed leg in all the diagrams signifies that all of these graphs are proportional to:

$$\mathrm{R}_{l_1 i_1 k_1}(-\mathbf{p}_1)\, h_{l_1}(p_1), \tag{C.1.1}$$

which means that third order one-loop corrections are only required to the zeroth order in the external momentum. For this reason we can simply consider the projector operators along the loop. Consider the first loop in the list of eighteen diagrams:

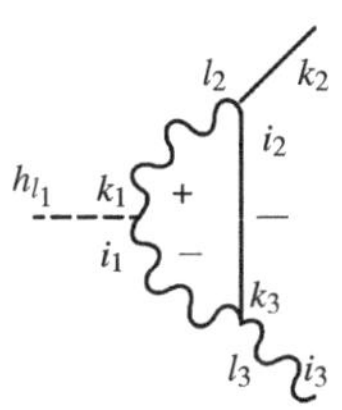

$$\propto \mathrm{R}_{l_2 i_2 k_2}(\mathbf{q})\, \mathrm{R}_{l_3 i_3 k_3}(\mathbf{q})\, \mathrm{P}_{i_2 k_3}(\mathbf{q})\, \delta_{i_1 l_3} \delta_{k_1 l_2}$$

Note that

$$q_{i_2} \mathrm{P}_{i_2 k_3}(\mathbf{q}) = 0, \tag{C.1.2}$$

so the contribution of this graph is zero. The same observation can be made explicit about the remaining seventeen graphs.

This is a tremendous simplification. We have to deal only with six diagrams. Consider the first graph:

$$\begin{aligned}\Psi_1 = &\left(-\frac{1}{2}\right) \int dp_1 dq_2 d\tilde{k}_3 \mathrm{R}_{l_1 i_1 k_1}(-\mathbf{p}_1)\, h^<_{l_1}(p_1)\, P^<_{i_2}\left(\tilde{k}_3\right) Q^<_{i_3}(q_2) \\ &\times \int dp_{23} dq_{13} d\tilde{k}_{12} \mathrm{R}_{l_2 i_2 k_2}\left(-\tilde{\mathbf{k}}_1\right) \mathrm{R}_{l_3 i_3 k_3}(-\mathbf{q}_1)\, |\mathbf{q}_3|^{-y}\, \mathrm{P}_{k_2 k_3}(\mathbf{q}_3)\, \delta_{k_1 l_2} \delta_{i_1 l_3} \\ &\times \delta(p_1 + p_2 + p_3)\, \delta(q_1 + q_2 + q_3)\, \delta\left(\tilde{k}_1 + \tilde{k}_2 + \tilde{k}_3\right) \delta\left(\tilde{k}_1 + p_3\right) \delta\left(\tilde{k}_2 + q_3\right) \delta(p_2 + q_1) \\ &\times C^-\left(\tilde{k}_1\right) G^-(\mathbf{q}_3)\, C^-(q_1).\end{aligned} \tag{C.1.3}$$

We proceed by integrating over momentum:

$$\begin{aligned}\Psi_1 = &\left(-\frac{1}{2}\right) \int dp_1 dq_2 d\tilde{k}_3 \mathrm{R}_{l_1 i_1 k_1}(-\mathbf{p}_1)\, h^<_{l_1}(p_1)\, P^<_{i_2}\left(\tilde{k}_3\right) Q^<_{i_3}(q_2)\, \delta\left(p_1 + q_2 + \tilde{k}_3\right) \\ &\times \int dq_3 \mathrm{R}_{l_2 i_2 k_2}\left(\tilde{\mathbf{k}}_3 - \mathbf{q}_3\right) \mathrm{R}_{l_3 i_3 k_3}\left(-\mathbf{p}_1 - \tilde{\mathbf{k}}_3 + \mathbf{q}_3\right) |\mathbf{q}_3|^{-y}\, \mathrm{P}_{k_2 k_3}(\mathbf{q}_3)\, \delta_{k_1 l_2} \delta_{i_1 l_3} \\ &\times C^-\left(-\tilde{k}_3 + q_3\right) G^-(q_3)\, C^-\left(p_1 + \tilde{k}_3 - q_3\right).\end{aligned} \tag{C.1.4}$$

The last two lines form the loop correction. Note, the contribution is of order $\mathbf{p}_1$, so we require a zero order correction in external momentum from the loop integral. This simplifies to:

$$\psi_1 = \int dq_3 \mathrm{R}_{l_2 i_2 k_2}(-\mathbf{q}_3)\, \mathrm{R}_{l_3 i_3 k_3}(\mathbf{q}_3)\, |\mathbf{q}_3|^{-y}\, \mathrm{P}_{k_2 k_3}(\mathbf{q}_3)\, \delta_{k_1 l_2} \delta_{i_1 l_3} C^-(q_3)\, G^-(q_3)\, C^-(-q_3)\,. \tag{C.1.5}$$

We can now contract the indices's to obtain:

$$\begin{aligned}\psi_1 &= -\int_{>} dq\, |\mathbf{q}|^{-y} q_{i_2} q_{i_3} \mathrm{P}_{k_1 k_2}(\mathbf{q})\, \mathrm{P}_{i_1 k_3}(\mathbf{q})\, \mathrm{P}_{k_2 k_3}(\mathbf{q})\, C^-(q)\, G^-(q)\, C^-(-q) \\ &= -\int_{>} dq\, |\mathbf{q}|^{-y} q_{i_2} q_{i_3} \mathrm{P}_{k_1 i_1}(\mathbf{q})\, C^-(q)\, G^-(q)\, C^-(-q)\,. \end{aligned} \tag{C.1.6}$$

Moreover, consider the frequency integral and the momentum integral separately:

$$\begin{aligned}\psi_1 &= -\int_{>} q^{d-1-y} dq \int \frac{d\Omega}{(2\pi)^d} q_{i_2} q_{i_3} \mathrm{P}_{k_1 i_1}(\mathbf{q}) \int \frac{d\omega}{2\pi} C^-(q)\, G^-(q)\, C^-(-q) \\ &= \int_{>} q^{d-3-y} dq \int \frac{d\Omega}{(2\pi)^d} q_{i_2} q_{i_3} q_{k_1} q_{i_1} \int \frac{d\omega}{2\pi} C^-(q)\, G^-(q)\, C^-(-q)\,. \end{aligned} \tag{C.1.7}$$

Finally, we need to calculate the frequency integral:

$$\begin{aligned}\int \frac{d\omega}{2\pi} C^-(q)\, G^-(q)\, C^-(-q) = &\int \frac{d\omega}{2\pi} \left(\frac{1}{i\omega + \nu q^2} - \frac{1}{i\omega + \mu q^2}\right)\left(\frac{1}{-i\omega + \nu q^2} - \frac{1}{-i\omega + \mu q^2}\right) \\ &\times \left(\frac{A}{\omega^2 + \nu^2 q^4} - \frac{B}{\omega^2 + \mu^2 q^4}\right). \end{aligned} \tag{C.1.8}$$

We make use of the identities:

$$\int \frac{dz}{2\pi} \frac{1}{\left(z^2 + \nu^2\right)^2} = \frac{1}{4\nu^3}, \tag{C.1.9}$$

$$\int \frac{dz}{2\pi} \frac{1}{\left(z^2 + \nu^2\right)\left(z^2 + \mu^2\right)} = \frac{1}{2\mu\nu(\nu + \mu)}, \tag{C.1.10}$$

$$\begin{aligned}\int \frac{dz}{2\pi} \frac{1}{(z - i\nu)(z + i\mu)\left(z^2 + \nu^2\right)} &= \int \frac{dz}{2\pi} \frac{1}{(z - i\nu)^2 (z + i\mu)(z + i\nu)} \\ &= \frac{1}{(\mu + \nu)^2 (\nu - \mu)} - \frac{1}{4\nu^2(\nu - \mu)}, \end{aligned} \tag{C.1.11}$$

We define the following quantity:

$$\begin{aligned}\mathscr{Y}_{+} =&A\left[\frac{1}{4\nu^3}+\frac{1}{2\mu\nu\left(\nu+\mu\right)}-\frac{2}{\left(\mu+\nu\right)^2\left(\nu-\mu\right)}+\frac{1}{2\nu^2\left(\nu-\mu\right)}\right]\\ &-B\left[\frac{1}{4\mu^3}+\frac{1}{2\mu\nu\left(\nu+\mu\right)}-\frac{2}{\left(\mu+\nu\right)^2\left(\mu-\nu\right)}+\frac{1}{2\mu^2\left(\mu-\nu\right)}\right].\end{aligned} \tag{C.1.12}$$

Thus,

$$\begin{aligned}\psi_1 =&\mathscr{Y}_{+}\int\limits_{>} q^{d-5-y}dq\int\frac{d\Omega}{(2\pi)^d}\frac{q_{i_2}q_{i_3}q_{k_1}q_{i_1}}{q^4}\\ =&\mathscr{Y}_{+}\Lambda^{\varepsilon}\delta\ell\frac{S_d}{(2\pi)^d}\frac{\left(\delta_{i_2i_3}\delta_{i_1k_1}+\delta_{i_2k_1}\delta_{i_3i_1}+\delta_{i_2i_1}\delta_{i_3k_1}\right)}{d\left(d+2\right)}.\end{aligned} \tag{C.1.13}$$

Subsequent analysis requires us to consider the remaining 5 graphs.

$$\begin{aligned}\Psi_2 =&\left(-\frac{1}{2}\right)\int dp_1dq_2d\tilde{k}_3\mathrm{R}_{l_1i_1k_1}\left(-\mathbf{p}_1\right)h^{<}_{l_1}\left(p_1\right)P^{<}_{i_2}\left(\tilde{k}_3\right)Q^{<}_{i_3}\left(q_2\right)\\ &\times\int dp_{23}dq_{13}d\tilde{k}_{12}\mathrm{R}_{l_2i_2k_2}\left(-\tilde{\mathbf{k}}_1\right)\mathrm{R}_{l_3i_3k_3}\left(-\mathbf{q}_1\right)|\mathbf{q}_3|^{-y}\,\mathrm{P}_{i_1k_3}\left(\mathbf{q}_3\right)\delta_{k_1l_2}\delta_{k_2l_3}\\ &\times\delta\left(p_1+p_2+p_3\right)\delta\left(q_1+q_2+q_3\right)\delta\left(\tilde{k}_1+\tilde{k}_2+\tilde{k}_3\right)\delta\left(\tilde{k}_1+p_3\right)\delta\left(p_2+q_3\right)\delta\left(\tilde{k}_2+q_1\right)\\ &\times C^{-}\left(\tilde{k}_1\right)G^{-}\left(q_3\right)C^{-}\left(q_1\right).\end{aligned} \tag{C.1.14}$$

Integrate over momentum:

$$\begin{aligned}\psi_2 =&\int dq_3\mathrm{R}_{l_2i_2k_2}\left(\tilde{\mathbf{k}}_3+\mathbf{q}_2+\mathbf{q}_3\right)\mathrm{R}_{l_3i_3k_3}\left(\mathbf{q}_2+\mathbf{q}_3\right)|\mathbf{q}_3|^{-y}\,\mathrm{P}_{i_1k_3}\left(\mathbf{q}_3\right)\delta_{k_1l_2}\delta_{k_2l_3}\\ &\times\delta\left(p_1+q_2+\tilde{k}_3\right)C^{-}\left(-\tilde{k}_3-q_2-q_3\right)G^{-}\left(\mathbf{q}_3\right)C^{-}\left(-q_2-q_3\right).\end{aligned} \tag{C.1.15}$$

Taking the limit of external momentum going to zero we obtain:

$$\psi_2 =\int dq_3\mathrm{R}_{l_2i_2k_2}\left(\mathbf{q}_3\right)\mathrm{R}_{l_3i_3k_3}\left(\mathbf{q}_3\right)|\mathbf{q}_3|^{-y}\,\mathrm{P}_{i_1k_3}\left(\mathbf{q}_3\right)\delta_{k_1l_2}\delta_{k_2l_3}C^{-}\left(-q_3\right)G^{-}\left(\mathbf{q}_3\right)C^{-}\left(-q_3\right). \tag{C.1.16}$$

This is then followed by index contraction:

$$\psi_2 =-\int dq\frac{q_{i_2}q_{i_3}q_{i_1}q_{k_1}}{q^2}|\mathbf{q}|^{-y}\,C^{-}\left(-q\right)G^{-}\left(\mathbf{q}\right)C^{-}\left(-q\right). \tag{C.1.17}$$

The frequency integral then reads:

$$\int \frac{d\omega}{2\pi} C^{-}(-q)\, G^{-}(\mathbf{q})\, C^{-}(-q) = \int \frac{d\omega}{2\pi} \left(\frac{1}{-i\omega + \nu q^2} - \frac{1}{-i\omega + \mu q^2} \right) \times \left(\frac{1}{-i\omega + \nu q^2} - \frac{1}{-i\omega + \mu q^2} \right) \times \left(\frac{A}{\omega^2 + \nu^2 q^4} - \frac{B}{\omega^2 + \mu^2 q^4} \right) \tag{C.1.18}$$

Evaluation of each integral frequency integral can be compactly written using the following quantity:

$$\mathscr{X} = A \left(\frac{1}{8\nu^3} - \frac{\mu}{2\nu^2 (\nu + \mu)^2} \right) - B \left(\frac{1}{8\mu^3} - \frac{\nu}{2\mu^2 (\nu + \mu)^2} \right). \tag{C.1.19}$$

Then, we can conclude:

$$\psi_2 = -\mathscr{X} \Lambda^{\varepsilon} \delta\ell \frac{S_d}{(2\pi)^d} \frac{\left(\delta_{i_2 i_3}\delta_{i_1 k_1} + \delta_{i_2 k_1}\delta_{i_3 i_1} + \delta_{i_2 i_1}\delta_{i_3 k_1}\right)}{d\,(d+2)}. \tag{C.1.20}$$

Next, we move onto the third contribution:

$$\begin{aligned}\Psi_3 = &\left(-\frac{1}{2}\right) \int dp_1 dq_2 d\tilde{k}_3 \mathrm{R}_{l_1 i_1 k_1}(-\mathbf{p}_1)\, h^{<}_{l_1}(p_1)\, P^{<}_{i_2}\left(\tilde{k}_3\right) Q^{<}_{i_3}(q_2) \\ &\times \int dp_{23} dq_{13} d\tilde{k}_{12} \mathrm{R}_{l_2 i_2 k_2}\left(-\tilde{\mathbf{k}}_1\right) \mathrm{R}_{l_3 i_3 k_3}(-\mathbf{q}_1)\, |\mathbf{p}_3|^{-y}\, \mathrm{P}_{k_1 k_2}(\mathbf{p}_3)\, \delta_{k_3 l_2} \delta_{i_1 l_3} \\ &\times \delta(p_1 + p_2 + p_3)\, \delta(q_1 + q_2 + q_3)\, \delta\left(\tilde{k}_1 + \tilde{k}_2 + \tilde{k}_3\right) \delta\left(p_3 + \tilde{k}_2\right) \delta\left(\tilde{k}_1 + q_3\right) \delta(p_2 + q_1) \\ &\times C^{-}\left(\tilde{k}_1\right) G^{-}(p_3)\, C^{-}(q_1).\end{aligned} \tag{C.1.21}$$

Integrate over the momentum:

$$\begin{aligned}\psi_3 = &\int dq_3 \mathrm{R}_{l_2 i_2 k_2}(\mathbf{q}_3)\, \mathrm{R}_{l_3 i_3 k_3}\left(-\mathbf{p}_1 - \tilde{\mathbf{k}}_3 + \mathbf{q}_3\right) \left|\tilde{\mathbf{k}}_3 - \mathbf{q}_3\right|^{-y} \mathrm{P}_{k_1 k_2}\left(\tilde{\mathbf{k}}_3 - \mathbf{q}_3\right) \delta_{k_3 l_2} \delta_{i_1 l_3} \\ &\times \delta\left(p_1 + q_2 + \tilde{k}_3\right) C^{-}(-q_3)\, G^{-}\left(\tilde{k}_3 - q_3\right) C^{-}\left(p_1 + \tilde{k}_3 - q_3\right).\end{aligned} \tag{C.1.22}$$

Take the limit of external momentum going to zero:

$$\psi_3 = \int dq \mathrm{R}_{l_2 i_2 k_2}(\mathbf{q})\, \mathrm{R}_{l_3 i_3 k_3}(\mathbf{q})\, |\mathbf{q}|^{-y}\, \mathrm{P}_{k_1 k_2}(\mathbf{q})\, \delta_{k_3 l_2} \delta_{i_1 l_3} C^{-}(-q)\, G^{-}(-q)\, C^{-}(-q). \tag{C.1.23}$$

We contract the indicies:

$$\psi_3 = \int dq q_{i_2} q_{i_3} \mathrm{P}_{i_1 k_1}(\mathbf{q}) |\mathbf{q}|^{-y} C^-(-q) G^-(-q) C^-(-q)$$
$$= -\int dq \frac{q_{i_2} q_{i_3} q_{i_1} q_{k_1}}{q^2} |\mathbf{q}|^{-y} C^-(-q) G^-(-q) C^-(-q). \quad \text{(C.1.24)}$$

This contribution is identical to ψ_2:

$$\psi_3 = -\mathscr{X} \Lambda^{\varepsilon} \delta \ell \frac{S_d}{(2\pi)^d} \frac{\left(\delta_{i_2 i_3}\delta_{i_1 k_1} + \delta_{i_2 k_1}\delta_{i_3 i_1} + \delta_{i_2 i_1}\delta_{i_3 k_1}\right)}{d(d+2)}. \quad \text{(C.1.25)}$$

Consider the fourth graph:

$$\Psi_4 = \left(-\frac{1}{2}\right) \int dp_1 dq_2 d\tilde{k}_3 \mathrm{R}_{l_1 i_1 k_1}(-\mathbf{p}_1) h_{l_1}^<(p_1) P_{i_2}^<\left(\tilde{k}_3\right) Q_{i_3}^<(q_2)$$
$$\times \int dp_{23} dq_{13} d\tilde{k}_{12} \mathrm{R}_{l_2 i_2 k_2}\left(-\tilde{\mathbf{k}}_1\right) \mathrm{R}_{l_3 i_3 k_3}(-\mathbf{q}_1) |\mathbf{p}_3|^{-y} \mathrm{P}_{k_1 k_3}(\mathbf{p}_3) \delta_{k_2 l_3} \delta_{i_1 l_2}$$
$$\times \delta(p_1 + p_2 + p_3) \delta(q_1 + q_2 + q_3) \delta\left(\tilde{k}_1 + \tilde{k}_2 + \tilde{k}_3\right) \delta(p_3 + q_3) \delta\left(\tilde{k}_2 + q_1\right) \delta\left(p_2 + \tilde{k}_1\right)$$
$$\times G^+(p_3) C^-(q_1) C^+\left(\tilde{k}_1\right). \quad \text{(C.1.26)}$$

Contract the momentum integrals:

$$\psi_4 = \int dq_3 \mathrm{R}_{l_2 i_2 k_2}(q_3 - p_1) \mathrm{R}_{l_3 i_3 k_3}\left(-p_1 + q_3 - \tilde{k}_3\right) |\mathbf{q}_3|^{-y} \mathrm{P}_{k_1 k_3}(\mathbf{q}_3) \delta_{k_2 l_3} \delta_{i_1 l_2}$$
$$\times \delta\left(p_1 + \tilde{k}_3 + q_2\right) G^+(-q_3) C^-\left(p_1 - q_3 + \tilde{k}_3\right) C^+(p_1 - q_3). \quad \text{(C.1.27)}$$

Take the limit of external momentum going to zero:

$$\psi_4 = \int dq q_{i_2} q_{i_3} \mathrm{P}_{i_1 k_1}(q) |\mathbf{q}|^{-y} C^-(-q) G^+(-q) C^+(-q)$$
$$= -\int dq \frac{q_{i_2} q_{i_3} q_{i_1} q_{k_1}}{q^2} |\mathbf{q}|^{-y} C^-(-q) G^+(-q) C^+(-q). \quad \text{(C.1.28)}$$

The frequency integral is:

$$\int \frac{d\omega}{2\pi} C^-(-q) G^+(-q) C^+(-q) = \int \frac{d\omega}{2\pi} \left(\frac{1}{-i\omega + \nu q^2} - \frac{1}{-i\omega + \mu q^2}\right)$$
$$\times \left(\frac{1}{-i\omega + \nu q^2} + \frac{1}{-i\omega + \mu q^2}\right) \times \left(\frac{A}{\omega^2 + \nu^2 q^4} + \frac{B}{\omega^2 + \mu^2 q^4}\right). \quad \text{(C.1.29)}$$

We define the quantity:

$$\mathscr{U}_{-}=\frac{A}{q^{6}}\left(\frac{1}{8\nu^{3}}-\frac{1}{2\nu\left(\nu+\mu\right)^{2}}\right)+\frac{B}{q^{6}}\left(\frac{1}{8\mu^{3}}-\frac{1}{2\mu\left(\nu+\mu\right)^{2}}\right), \tag{C.1.30}$$

We finally conclude:

$$\psi_{4}=-\mathscr{U}_{-}\delta\ell\frac{\Lambda^{\varepsilon}}{d\left(d+2\right)}\frac{S_{d}}{\left(2\pi\right)^{d}}\left(\delta_{i_{2}i_{3}}\delta_{i_{1}k_{1}}+\delta_{i_{2}k_{1}}\delta_{i_{3}i_{1}}+\delta_{i_{2}i_{1}}\delta_{i_{3}k_{1}}\right). \tag{C.1.31}$$

Second to last contribution:

$$\begin{aligned}\Psi_{5}=&\left(-\frac{1}{2}\right)\int dp_{1}dq_{2}d\tilde{k}_{3}\mathrm{R}_{l_{1}i_{1}k_{1}}\left(-\mathbf{p}_{1}\right)h_{l_{1}}^{<}\left(p_{1}\right)P_{i_{2}}^{<}\left(\tilde{k}_{3}\right)Q_{i_{3}}^{<}\left(q_{2}\right)\\&\times\int dp_{23}dq_{13}d\tilde{k}_{12}\mathrm{R}_{l_{2}i_{2}k_{2}}\left(-\tilde{\mathbf{k}}_{1}\right)\mathrm{R}_{l_{3}i_{3}k_{3}}\left(-\mathbf{q}_{1}\right)\left|\mathbf{q}_{3}\right|^{-y}\mathrm{P}_{k_{2}k_{3}}\left(\mathbf{q}_{3}\right)\delta_{k_{1}l_{3}}\delta_{i_{1}l_{2}}\\&\times\delta\left(p_{1}+p_{2}+p_{3}\right)\delta\left(q_{1}+q_{2}+q_{3}\right)\delta\left(\tilde{k}_{1}+\tilde{k}_{2}+\tilde{k}_{3}\right)\delta\left(p_{3}+q_{1}\right)\delta\left(\tilde{k}_{2}+q_{3}\right)\delta\left(p_{2}+\tilde{k}_{1}\right)\\&\times C^{+}\left(q_{1}\right)G^{-}\left(q_{3}\right)C^{+}\left(\tilde{k}_{1}\right).\end{aligned} \tag{C.1.32}$$

Momentum integration yields:

$$\begin{aligned}\psi_{5}=&\int dq_{3}\mathrm{R}_{l_{2}i_{2}k_{2}}\left(-\mathbf{q}_{3}+\tilde{\mathbf{k}}_{3}\right)\mathrm{R}_{l_{3}i_{3}k_{3}}\left(-\mathbf{p}_{1}+\mathbf{q}_{3}-\tilde{\mathbf{k}}_{3}\right)\left|\mathbf{q}_{3}\right|^{-y}\mathrm{P}_{k_{2}k_{3}}\left(\mathbf{q}_{3}\right)\delta_{k_{1}l_{3}}\delta_{i_{1}l_{2}}\\&\times\delta\left(p_{1}+q_{2}+\tilde{k}_{3}\right)C^{+}\left(p_{1}-q_{3}+\tilde{k}_{3}\right)G^{-}\left(q_{3}\right)C^{+}\left(q_{3}-\tilde{k}_{3}\right).\end{aligned} \tag{C.1.33}$$

Take the limit of external momentum going to zero:

$$\begin{aligned}\psi_{5}&=-\int dqq_{i_{2}}q_{i_{3}}\mathrm{P}_{i_{1}k_{1}}\left(\mathbf{q}\right)\left|\mathbf{q}\right|^{-y}C^{+}\left(-q\right)G^{-}\left(q\right)C^{+}\left(q\right)\\&=\int dq\frac{q_{i_{2}}q_{i_{3}}q_{i_{1}}q_{k_{1}}}{q^{2}}\left|\mathbf{q}\right|^{-y}C^{+}\left(-q\right)G^{-}\left(q\right)C^{+}\left(q\right).\end{aligned} \tag{C.1.34}$$

The frequency integral reads:

$$\begin{aligned}\int\frac{d\omega}{2\pi}C^{+}\left(-q\right)G^{-}\left(q\right)C^{+}\left(q\right)=&\int\frac{d\omega}{2\pi}\left(\frac{1}{-i\omega+\nu q^{2}}+\frac{1}{-i\omega+\mu q^{2}}\right)\\&\times\left(\frac{1}{i\omega+\nu q^{2}}+\frac{1}{i\omega+\mu q^{2}}\right)\times\left(\frac{A}{\omega^{2}+\nu^{2}q^{4}}-\frac{B}{\omega^{2}+\mu^{2}q^{4}}\right).\end{aligned} \tag{C.1.35}$$

Lets us identify this contribution by the following notation:

$$\begin{aligned}\mathscr{Y}_- =&A\left[\frac{1}{4\nu^3}+\frac{1}{2\mu\nu\left(\nu+\mu\right)}+\frac{2}{\left(\mu+\nu\right)^2\left(\nu-\mu\right)}-\frac{1}{2\nu^2\left(\nu-\mu\right)}\right]\\&-B\left[\frac{1}{4\mu^3}+\frac{1}{2\mu\nu\left(\nu+\mu\right)}+\frac{2}{\left(\mu+\nu\right)^2\left(\mu-\nu\right)}-\frac{1}{2\mu^2\left(\mu-\nu\right)}\right].\end{aligned} \tag{C.1.36}$$

The whole contribution then reads:

$$\begin{aligned}\psi_5 =&\mathscr{Y}_-\int dq\frac{q_{i_2}q_{i_3}q_{i_1}q_{k_1}}{q^4}\left|\mathbf{q}\right|^{-y-4}\\=&\mathscr{Y}_-\Lambda^{\varepsilon}\delta\ell\frac{S_d}{(2\pi)^d}\frac{\left(\delta_{i_2i_3}\delta_{i_1k_1}+\delta_{i_2k_1}\delta_{i_3i_1}+\delta_{i_2i_1}\delta_{i_3k_1}\right)}{d\left(d+2\right)}.\end{aligned} \tag{C.1.37}$$

At last we arrive at our final expression:

$$\begin{aligned}\Psi_6 =&\left(-\frac{1}{2}\right)\int dp_1dq_2d\tilde{k}_3\mathrm{R}_{l_1i_1k_1}\left(-\mathbf{p}_1\right)h^{<}_{l_1}\left(p_1\right)P^{<}_{i_2}\left(\tilde{k}_3\right)Q^{<}_{i_3}\left(q_2\right)\\&\times\int dp_{23}dq_{13}d\tilde{k}_{12}\mathrm{R}_{l_2i_2k_2}\left(-\tilde{\mathbf{k}}_1\right)\mathrm{R}_{l_3i_3k_3}\left(-\mathbf{q}_1\right)\left|\mathbf{p}_2\right|^{-y}\mathrm{P}_{k_2i_1}\left(\mathbf{p}_2\right)\delta_{k_1l_3}\delta_{k_3l_2}\\&\times\delta\left(p_1+p_2+p_3\right)\delta\left(q_1+q_2+q_3\right)\delta\left(\tilde{k}_1+\tilde{k}_2+\tilde{k}_3\right)\delta\left(p_3+q_1\right)\delta\left(q_3+\tilde{k}_1\right)\delta\left(\tilde{k}_2+p_2\right)\\&\times C^+\left(q_1\right)C^-\left(\tilde{k}_1\right)G^+\left(p_2\right).\end{aligned} \tag{C.1.38}$$

We proceed by contracting the momentum integrals:

$$\begin{aligned}\psi_6 =&\int dq_1\mathrm{R}_{l_2i_2k_2}\left(-\mathbf{q}_1-\mathbf{q}_2\right)\mathrm{R}_{l_3i_3k_3}\left(-\mathbf{q}_1\right)\left|\mathbf{q}_1-\mathbf{p}_1\right|^{-y}\mathrm{P}_{k_2i_1}\left(\mathbf{q}_1-\mathbf{p}_1\right)\delta_{k_1l_3}\delta_{k_3l_2}\\&\times\delta\left(p_1+q_2+\tilde{k}_3\right)C^+\left(q_1\right)C^-\left(q_1+q_2\right)G^+\left(q_1-p_1\right).\end{aligned} \tag{C.1.39}$$

We take the limit of external momentum going to zero:

$$\begin{aligned}\psi_6 =&\int dq\mathrm{R}_{l_2i_2k_2}\left(-\mathbf{q}\right)\mathrm{R}_{l_3i_3k_3}\left(-\mathbf{q}\right)\mathrm{P}_{k_2i_1}\left(\mathbf{q}\right)\left|\mathbf{q}\right|^{-y}\delta_{k_1l_3}\delta_{k_3l_2}C^+\left(q\right)C^-\left(q\right)G^+\left(q\right)\\=&-\int dq\frac{q_{i_2}q_{i_3}q_{i_1}q_{k_1}}{q^2}\left|\mathbf{q}\right|^{-y}C^+\left(q\right)C^-\left(q\right)G^+\left(q\right).\end{aligned} \tag{C.1.40}$$

The frequency integral reads:

$$\int \frac{d\omega}{2\pi} C^{+}(q)\, C^{-}(q)\, G^{+}(q) = \int \frac{d\omega}{2\pi} \left(\frac{1}{i\omega + \nu q^2} + \frac{1}{i\omega + \mu q^2}\right) \times \left(\frac{1}{i\omega + \nu q^2} - \frac{1}{i\omega + \mu q^2}\right) \times \left(\frac{A}{\omega^2 + \nu^2 q^4} + \frac{B}{\omega^2 + \mu^2 q^4}\right). \tag{C.1.41}$$

Denote a quantity:

$$\mathscr{U}_{+} = A\left(\frac{1}{8\nu^3} + \frac{1}{(\nu+\mu)^2\, 2\nu}\right) + B\left(\frac{1}{8\mu^3} + \frac{1}{(\nu+\mu)^2\, 2\mu}\right). \tag{C.1.42}$$

The end result reads:

$$\begin{aligned}\psi_6 =& -\mathscr{U}_{+} \int dq \frac{q_{i_2} q_{i_3} q_{i_1} q_{k_1}}{q^4} |\mathbf{q}|^{-y-4} \\ =& -\mathscr{U}_{+} \Lambda^{\varepsilon} \delta\ell \frac{S_d}{(2\pi)^d} \frac{\left(\delta_{i_2 i_3}\delta_{i_1 k_1} + \delta_{i_2 k_1}\delta_{i_3 i_1} + \delta_{i_2 i_1}\delta_{i_3 k_1}\right)}{d\,(d+2)}.\end{aligned} \tag{C.1.43}$$

Combining all these contributions from the frequency integrals we obtain the following factor:

$$\rightarrow A\left[\frac{1}{\mu\nu\,(\nu+\mu)} + \frac{\mu}{\nu^2\,(\nu+\mu)^2}\right] - B\left[\frac{1}{2\mu^3} + \frac{1}{\mu\nu\,(\nu+\mu)} + \frac{\nu}{\mu^2\,(\nu+\mu)^2}\right] \tag{C.1.44}$$

We now need to compute the numerical factors in front of this coefficient. A factor of six comes from the binomial expansion at a cubic order. All of the third order diagrams however have a prefactor of 1/3! from the cumulant expansion, which cancels the binomial factor of six. Third order terms all carry a minus sign as a consequence of a minus a half factor from the loop and a factor of λ^3. All together this gives a correction:

$$\begin{aligned}\delta\lambda =& -\frac{\lambda^3}{2}\left(A\left[\frac{1}{\mu\nu\,(\nu+\mu)} + \frac{\mu}{\nu^2\,(\nu+\mu)^2}\right] - B\left[\frac{1}{2\mu^3} + \frac{1}{\mu\nu\,(\nu+\mu)} + \frac{\nu}{\mu^2\,(\nu+\mu)^2}\right]\right) \\ &\times \Lambda^{\varepsilon} \delta\ell \frac{S_d}{(2\pi)^d} \frac{\left(\delta_{i_2 i_3}\delta_{i_1 k_1} + \delta_{i_2 k_1}\delta_{i_3 i_1} + \delta_{i_2 i_1}\delta_{i_3 k_1}\right)}{d\,(d+2)}.\end{aligned} \tag{C.1.45}$$

Further we note that a correction which is proportional to $\delta_{i_2 i_3}\delta_{i_1 k_1}$ vanishes, where as the last two combinations do not.

Correction to the vertex nPQ is identical. This can be seen first of all from the overall symmetry. Alternatively we see that by considering the non-vanishing diagrams which arise from the contraction of the following terms:

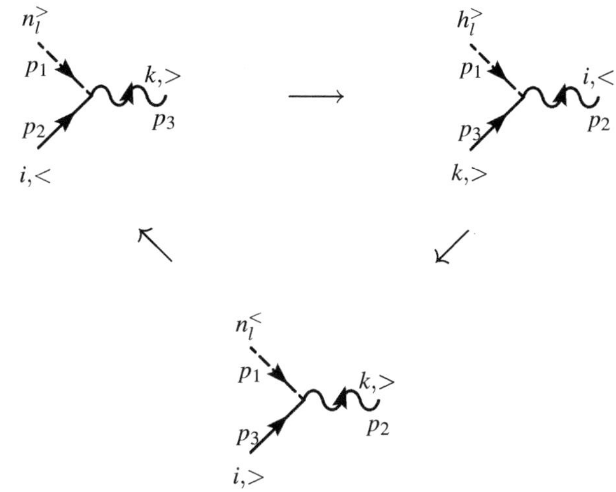

Effectively we only interchange the indicies $i_1 \leftrightarrow k_1$ in the bottom vertex. Since the loop is symmetrical under this interchange and this does not affect frequency integrals we can conclude that the correction is identical.

Appendix D
Transformation of the RG Corrections to Original Variables

The following transformations relate γ_+ and γ_- back to viscosity ν and magnetic diffusivity μ:

$$\nu = \gamma_+ + \gamma_-, \qquad \mu = \gamma_+ - \gamma_-. \tag{D.0.1}$$

D.1 Corrections to Viscosity

By collecting the results from Appendix C, we have:

$$\begin{aligned}
\delta\nu = {} & \frac{S_d}{(2\pi)^d}\delta\ell\Lambda^{\varepsilon}\Bigg[-\frac{(d+6-y)}{2d\,(d+2)}\tilde{\mathscr{B}}^0_- + \frac{d-2+y}{2d\,(d+2)}\tilde{\mathscr{B}}^0_+ + \frac{1}{d\,(d+2)}\left(\tilde{\mathscr{C}}^0_+ + \tilde{\mathscr{C}}^0_-\right) \\
& - \frac{(d+6-y)}{2d\,(d+2)}\tilde{\mathscr{B}}^0_+ + \frac{d-2+y}{2d\,(d+2)}\tilde{\mathscr{B}}^0_- + \frac{1}{d\,(d+2)}\left(\tilde{\mathscr{C}}^0_+ + \tilde{\mathscr{C}}^0_-\right) \\
& + \mathscr{B}^0_+\frac{3-d^2}{d\,(d+2)} - \mathscr{B}^0_-\frac{1}{d\,(d+2)} + \mathscr{B}^0_-\frac{3-d^2}{d\,(d+2)} - \mathscr{B}^0_+\frac{1}{d\,(d+2)}\Bigg] \\
& + \frac{S_d}{(2\pi)^d}\delta\ell\Lambda^{d-2}\Bigg[-\frac{(d+4-2y)}{2d\,(d+2)}\tilde{\mathscr{B}}^2_+ + \frac{d+2y}{2d\,(d+2)}\tilde{\mathscr{B}}^2_- \\
& + \frac{1}{d\,(d+2)}\left(\tilde{\mathscr{C}}^2_+ + \tilde{\mathscr{C}}^2_-\right) - \frac{(2+y)}{d\,(d+2)}\left(\tilde{\mathscr{B}}^2_+ + \tilde{\mathscr{B}}^2_-\right) - \frac{(d+4-2y)}{2d\,(d+2)}\tilde{\mathscr{B}}^2_- + \frac{d+2y}{2d\,(d+2)}\tilde{\mathscr{B}}^2_+ \\
& + \mathscr{B}^2_+\frac{3-d^2}{d\,(d+2)} - \mathscr{B}^2_-\frac{1}{d\,(d+2)} + \mathscr{B}^2_-\frac{3-d^2}{d\,(d+2)} - \mathscr{B}^2_+\frac{1}{d\,(d+2)} \\
& + \frac{1}{d\,(d+2)}\left(\tilde{\mathscr{C}}^2_+ + \tilde{\mathscr{C}}^2_-\right) - \frac{(2+y)}{d\,(d+2)}\left(\tilde{\mathscr{B}}^2_+ + \tilde{\mathscr{B}}^2_-\right)\Bigg].
\end{aligned} \tag{D.1.1}$$

Where we have split the loop contribution in the same way we have split noise covariance:

$$A = A_0 + A_2 k^{2+y}. \tag{D.1.2}$$

E. Barkhudarov, *Renormalization Group Analysis of Equilibrium and Non-equilibrium Charged Systems*, Springer Theses,
DOI: 10.1007/978-3-319-06154-2, © Springer International Publishing Switzerland 2014

Now, we exploit the definitions of frequency integrals:

$$\mathscr{B}_{+}(\Lambda)+\mathscr{B}_{-}(\Lambda)=\frac{1}{2}\left(\frac{A(\Lambda)}{\nu^{2}}+\frac{B(\Lambda)}{\mu^{2}}\right), \tag{D.1.3}$$

$$\tilde{\mathscr{B}}_{+}(\Lambda)+\tilde{\mathscr{B}}_{-}(\Lambda)=\frac{1}{2}\left(\frac{A(\Lambda)}{\nu^{2}}-\frac{B(\Lambda)}{\mu^{2}}\right), \tag{D.1.4}$$

$$\tilde{\mathscr{C}}_{+}(\Lambda)+\tilde{\mathscr{C}}_{-}(\Lambda)=\frac{3}{2}\left(\frac{A(\Lambda)}{\nu^{2}}-\frac{B(\Lambda)}{\mu^{2}}\right). \tag{D.1.5}$$

After a little bit of algebra we obtain:

$$\begin{aligned}\delta\nu=&\frac{S_d}{(2\pi)^d}\Lambda^{\varepsilon}\delta\ell\left\{\frac{A_0}{\nu^2}\frac{\left(y+4-d^2\right)}{2d\,(d+2)}-\frac{B_0}{\mu^2}\frac{\left(y+d^2\right)}{2d\,(d+2)}\right\}\\&+\frac{S_d}{(2\pi)^d}\Lambda^{d-2}\delta\ell\left(\frac{A_2}{\nu^2}+\frac{B_2}{\mu^2}\right)\frac{2-d^2}{2d\,(d+2)}.\end{aligned} \tag{D.1.6}$$

We can partially check the above result. Recall that the noise amplitude is:

$$A(k)=\left(A_0+A_2k^{2+y}\right)k^{-y}. \tag{D.1.7}$$

Therefore, by setting either $A_0=0$, or, $A_2=0$ and $y=-2$, we should obtain the same result for the correction to viscosity. Our result clearly satisfies this requirement. In addition it is in agreement with previous calculations for Navier-Stokes [5].

D.2 Corrections to Resistivity

Following the procedure we have performed to viscosity we can write:

$$\begin{aligned}\delta\mu=&\frac{S_d}{(2\pi)^d}\delta\ell\Lambda^{\varepsilon}\left[-\frac{(d+6-y)}{2d\,(d+2)}\tilde{\mathscr{B}}^0_{+}+\frac{d-2+y}{2d\,(d+2)}\tilde{\mathscr{B}}^0_{-}+\frac{1}{d\,(d+2)}\left(\tilde{\mathscr{C}}^0_{+}+\tilde{\mathscr{C}}^0_{-}\right)\right.\\&+\frac{(d+6-y)}{2d\,(d+2)}\tilde{\mathscr{B}}^0_{-}-\frac{d-2+y}{2d\,(d+2)}\tilde{\mathscr{B}}^0_{+}-\frac{1}{d\,(d+2)}\left(\tilde{\mathscr{C}}^0_{+}+\tilde{\mathscr{C}}^0_{-}\right)\\&\left.+\mathscr{B}^0_{+}\frac{3-d^2}{d\,(d+2)}-\mathscr{B}^0_{-}\frac{1}{d\,(d+2)}-\mathscr{B}^0_{-}\frac{3-d^2}{d\,(d+2)}+\mathscr{B}^0_{+}\frac{1}{d\,(d+2)}\right]\\&+\frac{S_d}{(2\pi)^d}\delta\ell\Lambda^{d-2}\left[-\frac{(d+4-2y)}{2d\,(d+2)}\tilde{\mathscr{B}}^2_{+}+\frac{d+2y}{2d\,(d+2)}\tilde{\mathscr{B}}^2_{-}\right.\\&+\frac{1}{d\,(d+2)}\left(\tilde{\mathscr{C}}^2_{+}+\tilde{\mathscr{C}}^2_{-}\right)-\frac{(2+y)}{d\,(d+2)}\left(\tilde{\mathscr{B}}^2_{+}+\tilde{\mathscr{B}}^2_{-}\right)+\frac{(d+4-2y)}{2d\,(d+2)}\tilde{\mathscr{B}}^2_{-}-\frac{d+2y}{2d\,(d+2)}\tilde{\mathscr{B}}^2_{+}\\&\mathscr{B}^2_{+}\frac{3-d^2}{d\,(d+2)}-\mathscr{B}^2_{-}\frac{1}{d\,(d+2)}-\mathscr{B}^2_{-}\frac{3-d^2}{d\,(d+2)}+\mathscr{B}^2_{+}\frac{1}{d\,(d+2)}\\&\left.-\frac{1}{d\,(d+2)}\left(\tilde{\mathscr{C}}^2_{+}+\tilde{\mathscr{C}}^2_{-}\right)+\frac{(2+y)}{d\,(d+2)}\left(\tilde{\mathscr{B}}^2_{+}+\tilde{\mathscr{B}}^2_{-}\right)\right].\end{aligned} \tag{D.2.1}$$

After some algebra we arrive at the following expression:

$$\delta\mu = \frac{S_d}{(2\pi)^d}\delta\ell\Lambda^{\varepsilon}\left[-\left(\tilde{\mathscr{B}}_+^0 - \tilde{\mathscr{B}}_-^0\right)\frac{1}{d} + \left(\mathscr{B}_+^0 - \mathscr{B}_-^0\right)\frac{2-d}{d}\right]$$
$$+ \frac{S_d}{(2\pi)^d}\delta\ell\Lambda^{d-2}\left[-\left(\tilde{\mathscr{B}}_+^2 - \tilde{\mathscr{B}}_-^2\right)\frac{1}{d} + \left(\mathscr{B}_+^2 - \mathscr{B}_-^2\right)\frac{2-d}{d}\right] \tag{D.2.2}$$

By combining the coefficients we obtain:

$$\mathscr{B}_+(\Lambda) - \mathscr{B}_-(\Lambda) = \frac{1}{\nu+\mu}\left(\frac{A}{\nu} + \frac{B}{\mu}\right), \tag{D.2.3}$$

$$\tilde{\mathscr{B}}_+(\Lambda) - \tilde{\mathscr{B}}_-(\Lambda) = \frac{1}{\nu+\mu}\left(\frac{A}{\nu} - \frac{B}{\mu}\right). \tag{D.2.4}$$

Thus, we get:

$$\delta\mu = \frac{S_d}{(2\pi)^d}\frac{\Lambda^{\varepsilon}}{(\nu+\mu)}\delta\ell\left\{\frac{A(\Lambda)}{\nu}\frac{(1-d)}{d} + \frac{B(\Lambda)}{\mu}\frac{(3-d)}{d}\right\}. \tag{D.2.5}$$

D.3 Corrections to the Noise Amplitudes

From our calculation we have derived that:

$$\delta(A(k)+B(k)) = k^2\left[\mathscr{A}_+(\Lambda)\frac{d^2-3}{d(d+2)} + \mathscr{A}_-(\Lambda)\frac{1}{d(d+2)}\right]\frac{S_d}{(2\pi)^d}\Lambda^{\tilde{\varepsilon}}\delta\ell, \tag{D.3.1}$$
$$\delta(A(k)-B(k)) = k^2\left[\mathscr{A}_-(\Lambda)\frac{d^2-3}{d(d+2)} + \mathscr{A}_+(\Lambda)\frac{1}{d(d+2)}\right]\frac{S_d}{(2\pi)^d}\Lambda^{\tilde{\varepsilon}}\delta\ell. \tag{D.3.2}$$

Let us make use of the following relations:

$$\mathscr{A}_-(\Lambda) + \mathscr{A}_+(\Lambda) = 2\frac{\left(A_0+\Lambda^{2+y}A_2\right)^2}{\nu^3} + 2\frac{\left(B_0+\Lambda^{2+y}B_2\right)^2}{\mu^3}, \tag{D.3.3}$$

$$\mathscr{A}_+(\Lambda) - \mathscr{A}_-(\Lambda) = \frac{4\left(A_0+\Lambda^{2+y}A_2\right)\left(B_0+\Lambda^{2+y}B_2\right)}{\nu\mu(\nu+\mu)}. \tag{D.3.4}$$

Then, it follows that corrections to the noise amplitudes are:

$$\delta A(k) = k^2\left(\frac{\left(A_0+\Lambda^{2+y}A_2\right)^2}{\nu^3}+\frac{\left(B_0+\Lambda^{2+y}B_2\right)^2}{\mu^3}\right)\frac{d^2-2}{d\,(d+2)}\frac{S_d}{(2\pi)^d}\Lambda^{\tilde{\varepsilon}}\delta\ell, \tag{D.3.5}$$

$$\delta B\,(k) = k^2\left(\frac{2\left(A_0+\Lambda^{2+y}A_2\right)\left(B_0+\Lambda^{2+y}B_2\right)}{\nu\mu\,(\nu+\mu)}\right)\frac{d-2}{d}\frac{S_d}{(2\pi)^d}\Lambda^{\tilde{\varepsilon}}\delta\ell. \tag{D.3.6}$$

More specifically

$$\delta A_2 = \left(\frac{A\,(\Lambda)^2}{\nu^3}+\frac{B\,(\Lambda)^2}{\mu^3}\right)\frac{d^2-2}{d\,(d+2)}\frac{S_d}{(2\pi)^d}\Lambda^{\tilde{\varepsilon}}\delta\ell, \tag{D.3.7}$$

$$\delta B_2 = \left(\frac{2A\,(\Lambda)\,B\,(\Lambda)}{\nu\mu\,(\nu+\mu)}\right)\frac{d-2}{d}\frac{S_d}{(2\pi)^d}\Lambda^{\tilde{\varepsilon}}\delta\ell, \tag{D.3.8}$$

where

$$A\,(\Lambda) \equiv A_0+\Lambda^{2+y}A_2, \tag{D.3.9}$$

$$B\,(\Lambda) \equiv B_0+\Lambda^{2+y}B_2. \tag{D.3.10}$$

D.4 Corrections of the Non-Linear Terms

To appreciate how non-linear terms are modified under RG we have to transform the field variables **P** and **Q** to **v** and **b**. The correction to λ is given by third order loop integral:

$$\begin{aligned}\delta\lambda_{i_1i_2i_3k_1} &= -\frac{\lambda^3}{2}\left(A\left[\frac{1}{\mu\nu\,(\nu+\mu)}+\frac{\mu}{\nu^2\,(\nu+\mu)^2}\right]-B\left[\frac{1}{2\mu^3}+\frac{1}{\mu\nu\,(\nu+\mu)}+\frac{\nu}{\mu^2\,(\nu+\mu)^2}\right]\right)\\ &\times\Lambda^{\varepsilon}\delta\ell\frac{S_d}{(2\pi)^d}\frac{\left(\delta_{i_2k_1}\delta_{i_3i_1}+\delta_{i_2i_1}\delta_{i_3k_1}\right)}{d\,(d+2)}.\end{aligned} \tag{D.4.1}$$

Consider now how this term appears in the action:

$$\begin{aligned}\rightarrow &\int\left(\lambda\mathrm{R}_{l_1i_1k_1}\,(-\mathbf{p})\,h^<_{l_1}\,(p)\,P^<_{i_1}\,(k)\,Q^<_{k_1}\,(q)+\delta\lambda_{i_1i_2i_3k_1}\mathrm{R}_{l_1i_1k_1}\,(-\mathbf{p})\,h^<_{l_1}\,(p)\right.\\ &\left.P^<_{i_2}\,(k)\,Q^<_{i_3}\,(q)\right)\delta\,(p+q+k)\\ + &\int\left(\lambda\mathrm{R}_{l_1k_1i_1}\,(-\mathbf{p})\,n^<_{l_1}\,(p)\,P^<_{i_1}\,(k)\,Q^<_{k_1}\,(q)+\delta\lambda_{i_1i_2i_3k_1}\mathrm{R}_{l_1k_1i_1}\,(-\mathbf{p})\,n^<_{l_1}\,(p)\right.\\ &\left.P^<_{i_2}\,(k)\,Q^<_{i_3}\,(q)\right)\delta\,(p+q+k),\end{aligned}$$

where the integration over all space-time variables is implied. To appreciate the meaning of this correction it is helpful to transform the field variables. Recall:

$$\mathbf{P} = \mathbf{v} + \mathbf{b}, \qquad \mathbf{Q} = \mathbf{v} - \mathbf{b}. \tag{D.4.2}$$

$$h = \xi^c + \psi^c, \tag{D.4.3}$$

$$n = \xi^c - \psi^c. \tag{D.4.4}$$

For compactness we define:

$$\delta\lambda_{i_1 i_2 i_3 k_1} = \delta\lambda \left(\delta_{i_2 k_1}\delta_{i_3 i_1} + \delta_{i_2 i_1}\delta_{i_3 k_1}\right). \tag{D.4.5}$$

Then, we can transform the above term:

$$\begin{aligned} \rightarrow \int 2\left(\lambda + 2\delta\lambda\right) \mathrm{R}_{l_1 i_1 k_1}\left(-\mathbf{p}\right) \xi^c_{l_1}\left(p\right) \left[v_{i_1}\left(k\right) v_{k_1}\left(q\right) - b_{i_1}\left(k\right) b_{k_1}\left(q\right)\right] \delta\left(p + q + k\right) \\ + \int 2\lambda \mathrm{R}_{l_1 k_1 i_1}\left(-\mathbf{p}\right) \psi^c_{l_1}\left(p\right) \left[b_{i_1}\left(k\right) v_{k_1}\left(q\right) - v_{i_1}\left(k\right) b_{k_1}\left(q\right)\right] \delta\left(p + q + k\right). \end{aligned} \tag{D.4.6}$$

From the above construct we can see that only non-linear terms in the Navier-Stokes equation are renormalized by $2\delta\lambda$. On the other hand, the non-linearity in the induction equation is not modified.

Appendix E
Rescaling and Differential RG Equations

E.1 Spatial and Temporal Transformations

The aim is to restore the cut-off featuring in the effective theory of large-scale interactions, namely $\Lambda b^{-1} \rightarrow \Lambda$, by means of temporal and spatial transformations:

$$\mathbf{p}' = b\mathbf{p}, \qquad \omega' = b^{z}\omega. \tag{E.1.1}$$

At this point we consider each term in the action separately. Consider the time derivative term:

$$i \int d\mathbf{p} d\omega h\left(-\mathbf{p}, -\omega\right) i\omega P\left(\mathbf{p}, \omega\right) \rightarrow b^{-d-2z+\xi+\eta} i \int d\mathbf{p}' d\omega' h' \left(-\mathbf{p}', -\omega'\right) i\omega' P' \left(\mathbf{p}', \omega'\right) \tag{E.1.2}$$

By convention we set this term to be invariant under such transformation because all of the theories governed by first order time derivative dynamics have the same form, namely:

$$\frac{\partial f}{\partial t}. \tag{E.1.3}$$

This means that the exponents have to satisfy the following equation:

$$d + 2z = \xi + \eta. \tag{E.1.4}$$

The next term is of the form:

$$i \int d\mathbf{p} d\omega h\left(-\mathbf{p}, -\omega\right) \gamma_{+} p^{2} P\left(\mathbf{p}, \omega\right) \rightarrow b^{-d-z+\xi+\eta-2} \gamma_{+} i \int d\mathbf{p}' d\omega' h' \left(-\mathbf{p}', -\omega'\right) p^{2} P' \left(\mathbf{p}', \omega'\right). \tag{E.1.5}$$

In terms of the way the coupling γ_{+} scales we can write:

$$\gamma_{+}' = b^{-d-z+\xi+\eta-2} \gamma_{+}. \tag{E.1.6}$$

E. Barkhudarov, *Renormalization Group Analysis of Equilibrium and Non-equilibrium Charged Systems*, Springer Theses,
DOI: 10.1007/978-3-319-06154-2, © Springer International Publishing Switzerland 2014

Using the limiting form and the identity we have established earlier this can written as follows:

$$\frac{d\gamma_{\pm}}{d\ell} = (z-2)\,\gamma_{\pm}. \tag{E.1.7}$$

Both couplings γ_+ and γ_- scale identically. The non-linear term scales in the following manner:

$$\lambda \quad \rightarrow \quad \lambda b^{-2d-2z+\xi-1+2\eta}. \tag{E.1.8}$$

Once again we can exploit the identity for the scaling exponents and consider the limit of $b \rightarrow 1$ to write:

$$\frac{d\lambda}{d\ell} = (\eta - d - 1)\,\lambda. \tag{E.1.9}$$

Finally, we have to consider how do the noise amplitudes scale. It is sufficient to consider a term of the form:

$$\int d\mathbf{p}d\omega h_i\,(-\mathbf{p},-\omega)\left(A_0 + A_2 p^{2+y}\right)|\mathbf{p}|^{-y}\,\mathrm{P}_{ij}\,(\mathbf{p})\,h_j\,(\mathbf{p},\omega)\,. \tag{E.1.10}$$

Under spatial and temporal transformations we obtain:

$$\rightarrow b^{-d-z+2\xi+y}\int d\mathbf{p}d\omega h_i\,(-\mathbf{p},-\omega)\left(A_0 + A_2 b^{-2-y} p^{2+y}\right)|\mathbf{p}|^{-y}\,\mathrm{P}_{ij}\,(\mathbf{p})\,h_j\,(\mathbf{p},\omega)\,. \tag{E.1.11}$$

Hence, we conclude:

$$\frac{dA_0}{d\ell} = (-d - z + 2\xi + y)\,A_0, \tag{E.1.12}$$

$$\frac{dA_2}{d\ell} = (-d - z + 2\xi - 2)\,A_2. \tag{E.1.13}$$

Evidently, it follows:

$$\frac{d\nu}{d\ell} = (z-2)\,\nu, \tag{E.1.14}$$

$$\frac{d\mu}{d\ell} = (z-2)\,\mu. \tag{E.1.15}$$

This scale transformation is now complemented by loop corrections we have calculated. This leads to the final set of equations:

$$\frac{d\nu}{d\ell} = \left(z - 2 + \frac{\lambda^2}{\nu} \left[\frac{A_0}{\nu^2} \left(\frac{d^2 - y - 4}{2d\,(d+2)} \right) + \frac{B_0}{\mu^2} \left(\frac{y + d^2}{2d\,(d+2)} \right) \right.\right.$$
$$\left.\left. + \frac{(d^2 - 2)}{2d\,(d+2)} \left(\frac{A_2}{\nu^2} + \frac{B_2}{\mu^2} \right) \Lambda^{d-2} \right] \frac{S_d}{(2\pi)^d} \Lambda^{\varepsilon} \right) \nu, \tag{E.1.16}$$

$$\frac{d\mu}{d\ell} = \left(z - 2 + \frac{\lambda^2 \Lambda^{\varepsilon}}{\mu\,(\nu + \mu)} \left[\frac{A}{\nu} \frac{(d-1)}{d} + \frac{B}{\mu} \frac{(d-3)}{d} \right] \frac{S_d}{(2\pi)^d} \right) \mu, \tag{E.1.17}$$

$$\frac{d\lambda}{d\ell} = \left(\eta - d - 1 - \frac{\lambda^2 \Lambda^{\varepsilon}}{d\,(d+2)} \frac{S_d}{(2\pi)^d} A \left[\frac{1}{\mu\nu\,(\nu + \mu)} + \frac{\mu}{\nu^2\,(\nu + \mu)^2} \right] \right. \tag{E.1.18}$$

$$\left. + \frac{\lambda^2 \Lambda^{\varepsilon}}{d\,(d+2)} \frac{S_d}{(2\pi)^d} B \left[\frac{1}{2\mu^3} + \frac{1}{\mu\nu\,(\nu + \mu)} + \frac{\nu}{\mu^2\,(\nu + \mu)^2} \right] \right) \lambda, \tag{E.1.19}$$

$$\frac{dA_0}{d\ell} = (-d - z + 2\xi + y)\, A_0, \tag{E.1.20}$$

$$\frac{dB_0}{d\ell} = (-d - z + 2\xi + y)\, B_0, \tag{E.1.21}$$

$$\frac{dA_2}{d\ell} = \left(-d - z + 2\xi - 2 + \frac{\lambda^2}{A_2} \frac{d^2 - 2}{d\,(d+2)} \left[\frac{A\,(\Lambda)^2}{\nu^3} + \frac{B\,(\Lambda)^2}{\mu^3} \right] \frac{S_d}{(2\pi)^d} \Lambda^{\tilde{\varepsilon}} \right) A_2, \tag{E.1.22}$$

$$\frac{dB_2}{d\ell} = \left(-d - z + 2\xi - 2 + \frac{\lambda^2}{B_2} \frac{d-2}{d} \left[\frac{2A\,(\Lambda)\, B\,(\Lambda)}{\nu\mu\,(\nu + \mu)} \right] \frac{S_d}{(2\pi)^d} \Lambda^{\tilde{\varepsilon}} \right) B_2. \tag{E.1.23}$$

We have used the following notation:

$$A\,(\Lambda) \equiv A_0 + \Lambda^{2+y} A_2, \qquad B\,(\Lambda) \equiv B_0 + \Lambda^{2+y} B_2, \tag{E.1.24}$$

and

$$\varepsilon = d - 4 - y \qquad \tilde{\varepsilon} = d - 2y - 6. \tag{E.1.25}$$

References

1. S.J. Camargo, H. Tasso, Renormalization group in magnetohydrodynamics turbulence, Phys. Fluids B 4(5) (1992)
2. A. Berera, S.R. Yoffe, Re-examination of the infrared properties of randomly stirred hydrodynamics. Phys. Rev. E **82**, 066304 (2010)
3. X.H. Wang, F. Wu, One modification to the Yakhot-Orszag calculation in the renormalization-group theory of turbulence. Phys. Rev. E **48**, R37 (1993)
4. M.K. Nandy, Symmetrization of the self-energy integral in the Yakhot-Orszag renormalization-group calculation. Phys. Rev. E **55**, 5455 (1997)
5. D. Forster, D.R. Nelson, M.J. Stephen, Large distance and long time properties of a randomly stirred fluid. Phys. Rev. A **16**, 732 (1977)

MIX
Papier aus verantwortungsvollen Quellen
Paper from responsible sources
FSC® C105338

If you have any concerns about our products,
you can contact us on
ProductSafety@springernature.com

In case Publisher is established outside the EU,
the EU authorized representative is:
Springer Nature Customer Service Center GmbH
Europaplatz 3, 69115 Heidelberg, Germany

Printed by Libri Plureos GmbH
in Hamburg, Germany